A series of student texts in

CONTEMPORARY BIOLOGY

General Editors:
Professor E. J. W. Barrington, F.R.S.
Professor Arthur J. Willis

The Mechanism of Photosynthesis

C. P. Whittingham
Ph.D.

Head of the Botany Department, Rothamsted Experimental Station;
Formerly Professor of Botany and of Plant Physiology,
Imperial College, University of London

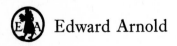 Edward Arnold

First published 1974
by Edward Arnold (Publishers) Limited
25 Hill Street, London W1X 8LL

Boards edition ISBN: 0 7131 2433 4
Paper edition   ISBN: 0 7131 2434 2

Printed in Great Britain by
William Clowes & Sons, Limited
London, Beccles and Colchester

Preface

The study of the mechanism of photosynthesis has attracted the interest of chemists, physicists and biologists. This book provides an introduction to present research in such a way that each specialist will be interested and led to appreciate the contribution of the others. The book does not attempt to consider all-the relevant literature but rather to provide a perspective for the university undergraduate and the young research worker.

Rothamsted 1973   C.P.W.

Contents

PREFACE   v

ABBREVIATIONS   viii

1 PHYSIOLOGY OF PHOTOSYNTHESIS   1
   The assimilatory quotient and the nature of the product of
      photosynthesis   3
   The diffusion of water vapour and carbon dioxide in leaves   5
   The rate of photosynthesis and the concentration of carbon dioxide   10
   Limiting factors   14

2 CARBON METABOLISM   16
   Introduction   16
   The discovery of the photosynthetic reduction cycle   18
   The reactions of the cycle   23
   Criticisms of the cycle   26

3 PHOTORESPIRATION   32
   The formation of glycollic acid during photosynthesis   32
   The compensation point   37
   The measurement of photorespiration   39
   The substrate of photorespiration   43
   The site of photorespiration   44

4 THE CONTRIBUTION OF THE CHLOROPLAST   47
   Carbon dioxide fixation by isolated chloroplasts   47
   The movement of carbon compounds between chloroplast
      and cytoplasm   51

5 EXCITATION AND FLUORESCENCE 56
Atomic absorption spectra 56
Molecular absorption spectra 58
Fluorescence and phosphorescence 59
The photosynthetic pigments *in vivo* 61
Fluorescence and energy transfer 66
Action spectra of photosynthesis 68

6 THE PHYSIOLOGICAL EVIDENCE FOR TWO
PHOTOCHEMICAL REACTIONS IN GREEN
PLANTS 71
Difference spectra 71
Enhancement spectra 77
Characterization of the two systems 79
Nuclear magnetic resonance spectra 81

7 THE COMPARATIVE BIOCHEMISTRY OF
PHOTOSYNTHESIS 84
Photosynthetic bacteria 84
Photosynthesis as an oxido-reduction reaction 86
The Hill reaction and ferredoxin 87
Photophosphorylation 90
The role of light energy in photosynthetic bacteria 92

8 ELECTRON TRANSPORT IN PHOTOSYNTHESIS 94
Cytochromes and photosynthesis 94
The separation of two photochemical systems 97
Carriers in electron transport 98
Physical separation of two particles 102
Mechanism of phosphorylation 103
Conformational changes in chloroplasts 105
Acid–base transfer and phosphorylation 106
The photosynthetic reaction centres 107

APPENDIX 109

GLOSSARY 113

REFERENCES 115

INDEX 123

Abbreviations

ADP	adenosine-5'-diphosphate
ATP	adenosine-5'-triphosphate
CoA	Coenzyme A
(CH_2O)	A generalized formula for sugars
DCMU	3-(3,4-dichlorophenyl)-1,1-dimethylurea
DPIP	2,6-dichlorophenol indophenol
EDTA	ethylene diamine tetra-acetate
INH	isonicotinylhydrazide (isoniazid)
NAD	nicotinamide adenine dinucleotide
NADP	nicotinamide adenine dinucleotide phosphate
P_i	inorganic phosphate
PGA	phosphoglyceric acid
PMS	N-methylphenazonium sulphate (Phenazine methosulphate)
PQ	plastoquinone
Q	a hypothetical quencher of fluorescence

I

Physiology of Photosynthesis

Photosynthesis is the process by which green plants provide our principal foodstuffs through the conversion of carbon dioxide and water to carbohydrates and oxygen. The reaction was first identified by the ability of the green leaf to produce oxygen, as shown by Joseph Priestley in 1771, very shortly after he had discovered the nature of the gas oxygen itself. Priestley demonstrated the complementary relationship between plants and animals by placing a mouse and a piece of mint together in a closed space under a bell jar. Even after one month the mouse survived without harm because the carbon dioxide produced by the breathing of the mouse was consumed during photosynthesis by the plant and converted back to oxygen. In later years Priestley was unable to repeat some of his earlier findings and it was Ingen-Housz who demonstrated that plants produce oxygen only when illuminated and that in darkness they behave like animals, consuming oxygen and producing carbon dioxide. Man is dependent on plants not only to supply food but to maintain his oxygen supply in the atmosphere. Similar experiments have been made in recent times to develop a biologically balanced system of man and plants on a microscale in a sealed submarine or spaceship.

By 1800 the basic metabolism of the green plant was understood. The process of photosynthesis producing food in the light was separated from that of respiration, in which foodstuffs were oxidized in the dark and which, overall, effected the opposite reaction. The process of photosynthesis requires a supply of energy whereas respiration releases energy. The energy contained within the organic carbon compounds produced in photosynthesis may be released either in the process of respiration or long after the plant has died by combustion of fossil fuels such as coal, petrol or oil. The green plant represents the most efficient energy converter known and converts solar energy into chemical energy. The radiation energy reaching the earth's surface from the sun is of the order of 2.0×10^{24} J per year, of which 6.0×10^{23} J is absorbed by plant vegetation. It has been estimated that 2–3% of the energy incident on, or 1 to 2% of the energy absorbed by, the plant is used in photosynthesis (say 1×10^{22} J), yet this suffices for the natural conversion of 3×10^{11} tonnes of carbon each year from the form of carbon dioxide to organic compounds, thus fixing approximately $10^{22}/3 \times 10^{11}$ J per tonne. According to the overall equation:

$$CO_2 + H_2O \longrightarrow (1/n)(CH_2O)_n + \tfrac{1}{2}O_2 + 472.8 \text{ kJ} \qquad \text{1.1}$$

for each 1 g of carbon in the form of CO_2 converted to the equivalent quantity (0.68 g) of carbohydrate, 472.8/12 or 40 kJ free energy are required. So that for each tonne of carbon (10^6 g) the theoretical requirement is 4.0×10^{10} J, in good agreement with the above estimate. The average annual yield of photosynthesis by crop plants has been well established as of the order of 5–10 g C/m²/day corresponding to 40 to 80 tonnes per hectare per year, or 16–32 English tons per acre.[143] According to equation 1.1 this would require an energy input of 2.0 to 4.0×10^6 J/m²/day. The energy falling on unit area of the earth's surface can be measured and in full sunlight the incident intensity may rise to 8.0 J/cm²/min, i.e. 6.0×10^7 J/m²/12 h day, although on dull days it will fall to one-third of this. Only the uppermost leaves of the plant will be exposed to the full intensity, many of the leaves lower down the plant being at least partially shaded although the total area of leaf per unit area of soil can rise in a crop plant at maturity to four or five. However, only part of the plant, i.e. the grain in a cereal plant, is harvested, so that the overall efficiency of approximately 0.5% is not unreasonable.

In the sea or in lakes and rivers, carbon dioxide is supplied to photosynthesis from dissolved salts of bicarbonate or carbonate in the medium bathing the plants. Most photosynthetic plants can use only the free carbon dioxide or undissociated carbonic acid, but some are capable of using bicarbonate ions as a direct source of carbon in photosynthesis. Terrestrial plants must obtain their carbon dioxide from the atmosphere and they do so largely through small holes in the epidermis of the leaf, the stomata. These holes permit not only the entry of carbon dioxide but also the escape of water vapour from the wet walls of the cells inside the leaf. The resulting loss of water by the plant in the process of transpiration is an inevitable consequence of the need by the leaf to absorb carbon dioxide. When the stomata are completely closed, a supply of carbon dioxide will be available for photosynthesis from the oxidation of compounds within the plant in the process of respiration and from the relatively slow diffusion through the waxy layer (cuticle) which covers the leaf surface.

In the broadest sense all the organic materials of the green plant may be considered to be the products of photosynthesis. In photosynthesis carbon taken up in the form of carbon dioxide from the air (or from solution in water) is incorporated into organic carbon compounds, and these are the starting point for all other biosyntheses (Fig. 1.1). The incorporation of

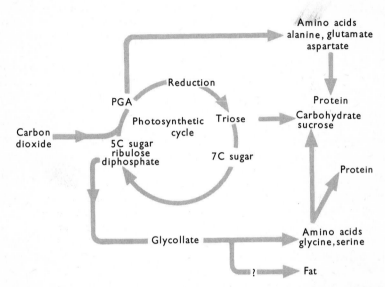

Fig. 1.1   Diagram to show the flow of carbon through the central photosynthetic cycle to a diversity of primary products.

carbon dioxide involves a reduction reaction which is balanced by a corresponding oxidation reaction in which oxygen is liberated from water. It is not always easy to distinguish clearly between the organic products formed directly by incorporating the carbon dioxide taken up in photosynthesis and those which are the result of subsequent metabolic transformations. But it is desirable that this distinction should be made.

The assimilatory quotient and the nature of the product of photosynthesis

The earliest observations on photosynthesis included quantitative determinations of the amount of carbon dioxide taken up relative to that of oxygen liberated as a reaction product. Recent experiments have shown that under a wide range of conditions, provided the oxygen concentration was low, the volume of oxygen liberated was almost equal to that of carbon dioxide absorbed (Fock et al.[42]); when the oxygen concentration was high the O_2 evolved was less than the carbon dioxide consumed. Using water distinguished by isotopic 'labelling' with ^{18}O it could be shown that almost all the oxygen produced in photosynthesis came from water (Dole and Jenks[32]). Thus the following equation is a more correct representation of photosynthesis than equation 1.1:

$$nCO_2 + 2nH_2O^* \longrightarrow nCH_2O + nH_2O + nO_2^* \qquad 1.2$$

showing that water is formed as a product of the reaction and contains oxygen which was originally present in the carbon dioxide. The organic product, denoted by CH_2O, has an empirical formula in which the elements carbon, hydrogen and oxygen are present in the proportion in carbohydrate. The production of protein or of fat would require that relatively less carbon dioxide should be consumed for the same production of oxygen; the formation of organic acids would relatively require more. Thus, it is still possible to observe an overall ratio of one carbon dioxide to one oxygen molecule if approximately equal quantities of organic acids and protein are formed.

Some investigations have analysed the gross composition of photosynthetic tissues after exposure to light or to dark, and attempted to determine a difference between them. Some of the commonest constituents of all plants, the simple carbohydrates, including sucrose and polysaccharides, such as starch, were found to increase in the light and decrease in the dark. The starch in green leaves is present in the subcellular structures containing the photosynthetic pigments, the chloroplasts, suggesting a close connection between starch formation and photosynthesis. Green leaves have a high protein content and in variegated forms the green areas contain more nitrogen than the less green. Analysis of isolated chloroplasts shows that they are unusually rich in protein and the protein is different in type from cytoplasmic protein. However, whether a leaf produces more protein in the light than it does in the dark is not clear from the evidence. In many photosynthetic organisms, especially in some algae, there is a high content of fat; but again the existence of a significant difference in content between light and dark grown cells has not been established.

One approach to the problem of the nature of the primary product of photosynthesis has been to measure the increase in the heat of combustion of a given area of leaf consequent upon a period of photosynthesis. In general it is found that the increase in heat of combustion per gram of carbon dioxide absorbed (10.8 kJ/g) is about intermediate between that calculated for sucrose or glucose synthesis (approximately 11 kJ/g CO_2 as in equation 1.1) and that for protein synthesis (1 g CO_2 forms about 0.5 g protein $+12.6$ kJ/g) but it is considerably less than that required for fat synthesis (1 g CO_2 forms about 0.4 g fat $+15.1$ kJ/g). The clearest evidence that protein could be formed during photosynthesis came from observations by Burström[20] using detached wheat leaves. Young leaves containing various quantities of nitrate were allowed to photosynthesize at low and at high intensities of light. Leaves containing little nitrate when they were illuminated at low intensities converted all the carbon dioxide absorbed into sugar. If the nitrate content was higher, more carbon dioxide was absorbed and the excess was used to convert the nitrogen to protein. The two kinds of product were additive and did not compete with one another at high light intensity, so that the total assimilation of carbon dioxide was greater and the formation of sugar relatively diminished in the leaves with higher nitrate content. This work suggested a relationship between nitrate assimilation and the uptake of carbon in photosynthesis, but it did not establish that the relationship was necessarily a direct one. By contrast, J. H. C. Smith[121] showed with sunflower leaves that almost the whole of the carbon dioxide absorbed during photosynthesis could be accounted for by the formation of carbohydrate. Many workers concluded that carbohydrate was the only primary product of photosynthesis and this view was supported by the observation that albino maize plants could be grown almost to maturity on sucrose as their only source of carbon.

The diffusion of water vapour and carbon dioxide in leaves

In the terrestrial plant which consumes carbon dioxide during photosynthesis, there must be a fall in concentration of carbon dioxide between the ambient air and the place of reaction in the chloroplasts of the leaf cell. Since carbon dioxide enters the leaf by a process of diffusion it moves from a position of higher to one of lower concentration. From the outside carbon dioxide diffuses in a gaseous phase through the stomata and the intercellular spaces of the leaf, then in an aqueous phase through the cell wall and within the cell up to the chloroplast. Since the cell walls of the mesophyll cells are wet the air immediately adjacent to them in the intercellular spaces will be very nearly saturated with water vapour at the temperature of the leaf. Under almost all conditions the atmosphere outside the leaf will not be saturated so that water will diffuse out from the mesophyll cells through the air spaces of the leaf and beyond the leaf surface into the atmosphere. The rate of exchange of carbon dioxide inwards or of water vapour outwards between the leaves and the surrounding atmosphere is given according to Fick's Law for transfer by molecular diffusion, that is the rate of flow is related to the difference in concentration between the supply and the point of consumption and inversely to the distance separating them. The quantity

S flowing per unit time t, when the difference in concentration is δc and the distance apart of source and sink is δx, is expressed by:

$$\frac{\delta S}{\delta t} = -KA\frac{\delta c}{\delta x}$$

This proportionality involves a coefficient of diffusion, K, whose dimensions are cm^2/s if S is in cm^3, A in cm^2, x in cm and δc cm^3/cm^3; its value varies according to the nature of the substance diffusing and the medium through which it diffuses. When this equation is applied to the process of transpiration we assume that the rate of transpiration is proportional to the difference in vapour pressure immediately adjacent to the wet cell wall within the leaf and that in the outside atmosphere above the leaf, an area term A and a length term δx. From determinations of the rate of transpiration under measured conditions of relative humidity, it is possible to derive from this expression a value for $K/\delta x$, provided A is known. The early workers Brown and Escombe[17] assumed that A could be taken as the total area of stomatal pore*, and δx the average length. Then substituting the known value for K, the coefficient of diffusion of water vapour in air, they calculated the rate of transpiration. The calculated rate proved to be 3–5 times larger than the actual value measured for a sunflower leaf in the field on a bright sunny day. It follows that the diffusion path for transpiration must be greater than that assumed and include additional components which account for the total distance from the wet mesophyll cell wall up to the leaf surface and beyond, including a layer of air on the surface of the leaf called the 'shearing' layer. Three components of the transpiration path have been identified:

1. The path inside the leaf starting at the wet mesophyll cell wall and extending up to the stoma r_m,
2. the stomatal path r_s and
3. the depth of still air upon the leaf surface, the 'shearing' layer r_a.

The rate of transpiration can then be written as:

$$T = \frac{K[(H_2O)\text{ leaf} - (H_2O)\text{ air}]}{r_a + r_s + r_m}$$

where $(H_2O)_a$ represents the concentration of water vapour at a. Note that the symbols r_a etc. are sometimes expressed relative to the diffusion constant, i.e. as $R_a = r_a/K$ with the dimensions of s/cm. Brown and Escombe's observations suggest that the stomatal path r_s can account for only about one-third to one-fifth of the total path, i.e. $r_a + r_s + r_m$.

* They ignored diffusion through the waxy layer on the leaf surface, i.e. cuticular transpiration, which in many plants probably accounts for about 10% of the total flow.

There is a thin layer of air adjacent to the leaf surface which is relatively stationary; this is referred to as the boundary or 'shearing' layer, mentioned above, and represents a stationary region where air adheres to the leaf surface. Diffusion between the leaf and the external air must take place through this layer. When the wind speed above the leaf is relatively slow the layer is thick and diffusion of water vapour or carbon dioxide across it correspondingly slow; but increase in wind velocity will decrease the depth of the 'shearing' layer increasing the rate of flow and hence the rate of transpiration. There is an empirical relationship between the depth of the 'shearing' layer and the velocity of the wind across the leaf surface,[101] namely

$$R_a = 1.3 \sqrt{\frac{I}{u}},$$

where I = leaf width (cm), 1.3 = constant ($s^{\frac{1}{2}}/cm$),
$\quad u$ = wind speed (cm/s).

It will be noted the depth of the layer is also a function of leaf size because the 'shearing' layer is smaller at the edge of the leaf than towards the centre. For a wind speed of 4 m.p.h. or 200 cm/s and a leaf 2 cm wide

$$R_a = 1.3 \sqrt{\frac{2}{200}} = 0.13 \text{ s/cm} \quad \text{or} \quad r_a = 0.03 \text{ cm}.$$

It has been calculated that normally on a bright sunny day with only a gentle breeze the shearing layer accounts for between $\frac{1}{2}$ and $\frac{1}{4}$ of the total transpiration path. It follows that under these conditions, changes in the stomatal aperture, due for example to changes in light intensity, water regime, or in carbon dioxide partial pressure surrounding the leaf, are not going to have very significant effects upon the rate of transpiration.

Similar considerations apply to photosynthesis. The average rate of photosynthesis for a sunflower leaf under good conditions is of the order of 100 $mm^3 CO_2/cm^2/h$, or $3.3 \times 10^{-5} cm^3 CO_2/cm^2/s$. From this we can deduce a value for $\sum R = \delta x/K$ (per unit area of leaf surface) of the order of 10 s/cm,

for $P = K.A \dfrac{\delta c}{\delta x}$, $\quad$ i.e. $\dfrac{\delta x}{K} = \dfrac{\delta c}{P/A}$.

Since $\delta c = (CO_2)_{air} - (CO_2)_{chloroplast} = 3 \times 10^{-4} cm^3/cm^3$,

when $(CO_2)_{air} = 0.03\% = 3 \times 10^{-4} cm^3/cm^3$

and $(CO_2)_{chloroplast} = 0$ (assumed),

therefore $\sum R = \dfrac{3 \times 10^{-4}}{3.3 \times 10^{-5}} \simeq 9$ s/cm.

From the known value of the coefficient of diffusion K for CO_2 diffusing in air, $K_{CO_2}^{air} = 0.14$ cm^2/s, it follows that the actual physical length of air diffusion path equivalent to the complex leaf structure is of the order of 1.5 cm. If the rate of photosynthesis is double that given above, then $\delta x/K$ becomes of the order of 5 s/cm and the equivalent diffusing path is approximately 0.75 cm.

For a given leaf the simplest assumption is that the equivalent diffusion path for transpiration is identical with that for photosynthesis up to the mesophyll cell wall; it follows that the rate of transpiration should bear some relationship to the rate of photosynthesis and is given in the simplest case by the following expression:

$$\frac{T_{\text{Transpiration}}}{P_{\text{Photosynthesis}}} = \frac{K_1(H_2O' - H_2O'')}{K_2(CO_2'' - CO_2')} \simeq 1.85 \frac{[H_2O' - H_2O'']}{[CO_2'' - CO_2']}$$

when K_1 is the coefficient of diffusion for water vapour in air, $H_2O' - H_2O''$ represents the difference in vapour pressure inside and outside the leaf, K_2 is the coefficient of diffusion for carbon dioxide in air and $CO_2'' - CO_2'$ the equivalent concentration gradient. At 20°C $K_1 = 0.257$ cm^2/s and $K_2 = 0.14$ cm^2/s. Penman and Schofield[112] found that turf during the summer lost 11 000 tonnes of water per 0.405 hectares and produced 2040 kilogrammes of dry matter. Values of the transpiration ratio, i.e. g water lost/g carbohydrate produced, range from 150 to 1500 so that the parameter is determined by many variables. Maize leaves produce a relatively high ratio of dry matter per water loss and during a day with bright illumination and high temperature lose 35 mg H_2O per mg CO_2 fixed.[19]

By use of such an expression we can calculate the concentration of carbon dioxide $[CO_2']$ at a distance along the diffusion path equal to the length of diffusion path for transpiration. This will provide an indication of the concentration of carbon dioxide near the mesophyll cell wall; it can be calculated as of the order of $\frac{5}{6}$ths that of the concentration in the external atmosphere. In the process of photosynthesis the total diffusion path is, however, longer and more complex than for transpiration; since not only does it include the three terms already discussed for transpiration but it must include also diffusion through the wet mesophyll cell wall up to the chloroplast surface and further into the chloroplast to the point at which it reacts. Approximately $\frac{2}{3}$rds to $\frac{5}{6}$ths of the equivalent air diffusion path for photosynthesis is internal to the mesophyll cell wall. Possibly $\frac{1}{3}$rd of this distance is due to penetration through the wet mesophyll cell wall, since the coefficient of diffusion for carbon dioxide through water is 1/10 000 that through air. Consequently, the physical distance through the *wet* cell wall assumes a much greater magnitude when it is converted to an equivalent *air* diffusion path. The remaining $\frac{2}{3}$rds, i.e. about $\frac{4}{9}$ths of the total path, is within the mesophyll cell. The average distance which a molecule of carbon dioxide must diffuse within the chloroplast for certainty of reaction i.e. the distance to the point at which the average concentration is decreased to zero, is referred to as the reaction or carboxylation resistance.

Since the stomatal path is an even smaller fraction of the total diffusion path for photosynthesis than that for transpiration, changes in stomatal aperture have even less effect on photosynthesis than on transpiration. It is only when the stomata close appreciably and their contribution to the total diffusion path becomes relatively large that they exert a significant influence on the rate of transpiration; even greater closure is required before they can exert a significant effect on the rate of photosynthesis. This can be demonstrated by the application of chemical inhibitors which primarily affect the stomatal mechanism. For example, Slatyer and Bierhuizen[120] showed that the stomata of cotton leaves sprayed with 10^{-4}M phenyl mercuric acetate remained closed for many days but growth of the plants

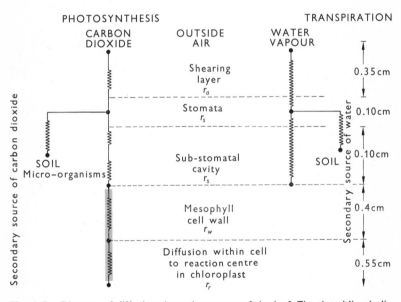

Fig. 1.2   Diagram of diffusion through stomata of the leaf. The tinted line indicates diffusion through the aqueous phase. The values on the right hand side give the approximate equivalent air diffusion path in cm for a leaf on a sunny day with relatively still air. According to the velocity of the wind r_a may vary from 0.50 to zero, according to stomatal aperture r_s from 0.1 to 1.0 and, according to reaction conditions, r_r from 0.4 to 1.0. The rate of photosynthesis observed will be in the ratio of 0.1 : 1.5, i.e. 1 : 15 to that predicted if the stomata constituted the total diffusion path and the rate of transpiration 0.1 : 0.55, i.e. 1 : 5.5. Brown and Escombe[17] calculated from their data ratios of 1 : 16 and 1 : 5 respectively.

was not greatly affected; transpiration was decreased 40% but photosynthesis only 10%.

When such considerations are applied to a crop growing in the field, other factors complicate the situation; for example, there is a significant contribution of carbon dioxide (possibly up to one-third of the total) from the activity of micro-organisms in the soil so that an additional supply must be entered in parallel to the supply from the external atmosphere. Similarly in considerations of transpiration, there is an additional supply of water vapour from the soil to the leaf in parallel with the supply from the air above the leaf. These relationships have been formalized by Penman and are shown diagrammatically in Figure 1.2.

The rate of photosynthesis and the concentration of carbon dioxide

Thus the concentration of carbon dioxide outside the leaf surface, and that available to the photosynthetic reaction centre in the chloroplast, do not bear a simple relationship to each other. Differences in structure and of internal anatomy between different types of leaf may have significant effects on the relationships between these two quantities. In Chapter 2 reference will be made to tropical grass species the leaves of which have large intercellular spaces between the mesophyll cells which have relatively few chloroplasts and the closely packed cells of the bundle sheath which are densely packed with chloroplasts arranged around the veins (Fig 1.3); both these factors may increase the equivalent diffusion resistance of the leaf for photosynthesis. It follows that the relationship between the rate of photosynthesis and the concentration of carbon dioxide in the outside atmosphere will vary in leaves of different species and under different environmental conditions. To deduce from the relationship observed between the rate of photosynthesis and the concentration of carbon dioxide in the atmosphere the true relationship between the rate of photosynthesis and the concentration at the reaction centre within the chloroplast involves complex calculation. Failure to appreciate this point led to considerable confusion between 1920 and 1936 when different investigators sought to maintain that a particular relationship observed by them experimentally had greater validity than that of some other observer. For this reason investigators later turned away from the study of such effects in leaves and preferred to investigate the kinetics of photosynthesis in an aqueous suspension of unicellular algae.

Even in the case of a uniform suspension of small microscopic cells, such as of the alga *Chlorella*, there must be some difference between the concentration of carbon dioxide in the bulk of the medium and that immediately adjacent to the chloroplast surface. The diffusion equation given earlier in this chapter can be applied, remembering that the diffusion path

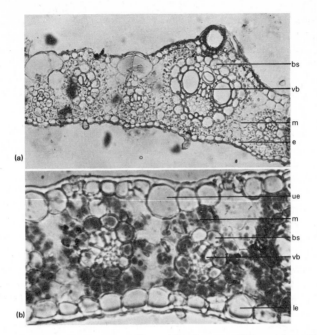

Fig. 1.3   T.S. leaf of (a) Sugar Cane (*Saccharum officinarum*), ×300, and (b) Napier Grass (*Pennisetum purpureum*), ×450. bs, bundle sheath; vb, vascular bundle; m, mesophyll; e, epidermis; le, lower epidermis; ue, upper epidermis. (Photographs by courtesy of Dr. Chris Bucke and Dr. J. Coombs.)

for a sphere reacting uniformly over its surface is equal to the radius of the sphere. Fortunately in this case we are not concerned with complications due to stomata whose aperture can vary with the concentration of carbon dioxide. From a knowledge of the rate of reaction per unit surface area, the concentration in the external medium, and the radius of the sphere, the concentration difference between the outside medium and the chloroplast of the *Chlorella* cell can be calculated. The percentage difference in concentration is of the order of 30%. The relationship between the rate of photosynthesis and the concentration of carbon dioxide for a suspension of *Chlorella* approximates to a rectangular hyperbola,

i.e.
$$P = P_\infty \frac{C}{C+k_c}$$

where P=rate of photosynthesis, C=concentration of carbon dioxide, P_∞ is the rate of photosynthesis when the concentration is very high and k_c=constant.

This is the type of relationship observed between the rate of an enzyme-catalysed reaction and the substrate concentration in a homogenous system. The constant k_c is inversely proportional to the affinity of the carboxylating enzyme in photosynthesis; in *Chlorella* k_c is approximately 5×10^{-6}M carbon dioxide.

It is clearly important to know whether this constant is the same for all photosynthetic systems. To do this we must attempt to deduce, from the observed relationship between the rate of photosynthesis of a leaf and the concentration of carbon dioxide in the external air, the actual relationship between the rate and the concentration of carbon dioxide at the reaction centre within the chloroplast. If it is assumed that the relationship between the internal and external concentration is given by the diffusion equation, and the relationship between rate and concentration at the reaction centre is that characteristic for simple enzyme systems, it can be deduced that the relationship between the rate of the reaction and the concentration in the external phase should be quadratic.

For $P = P_\infty (C/C + k_c)$ when $C =$ concentration of carbon dioxide at the chloroplast and the rate of supply by diffusion is $P = (K/\delta x) (C_{air} - C) + r$, r representing some contribution from the respiratory activity of the leaf. In a steady state these two rates must be equal, whence from the second equation

$$C_{air} - C = \frac{\delta x (P - r)}{K} \quad \text{or} \quad C = C_{air} - \frac{\delta x}{K} (P - r)$$

Substituting this in the first equation

$$P = P_\infty \cdot \frac{C_{air} - (\delta x / K)(P - r)}{C_{air} - (\delta x / K)(P - r) + k_c}$$

i.e. $\quad \dfrac{\delta x}{K} P^2 - P \left[r \dfrac{\delta x}{K} + C_{air} + k_c + \dfrac{\delta x}{K} P_\infty \right] + P_\infty C_{air} + \dfrac{\delta x}{K} r P_\infty = 0$

or $\quad P^2 - P \left[r + P_\infty + \dfrac{K}{\delta x} (C_{air} + k_c) \right] + P_\infty \left(C_{air} \dfrac{K}{\delta x} + r \right) = 0.$

Provided the diffusion path becomes very large, i.e. $Kk_c/\delta x$ becomes small, this expansion simplifies into an expression for two straight lines

i.e. $\quad P^2 - P \left[r + P_\infty + \dfrac{K}{\delta x} C_{air} \right] + P_\infty r + P_\infty C_{air} \dfrac{K}{\delta x} = 0.$

i.e. $\quad (P - P_\infty)(P - \dfrac{K}{\delta x} C_{air} - r) = 0$

or/either $\quad P = P_\infty \quad \text{or} \quad P = \dfrac{K}{\delta x} C_{air} + r;$

i.e. at lower concentrations there is a linear relationship between rate and the air concentration and at higher concentrations the rate is constant and independent of concentration and determined solely by the maximal rate of the enzyme system involved. Such a relationship might be deduced from first principles. When the concentration around the reaction centre is very high the system is working at maximal rate, i.e. $P = P_\infty$, and changes tending to increase the supply, such as further raising the concentration in the external environment or decreasing the distance between outside and inside, do not result in any increase in rate. All that will happen is that the concentration around the reaction centre will rise. If a factor, e.g. temperature, is changed so that the maximal velocity of the reaction P_∞ is increased, the rate of consumption will increase and the concentration around the centre will fall until the rate of supply again equals consumption. On the other hand when the concentration around the centre is such that the rate of consumption is less than maximal, i.e. $P = (K/\delta x)C_{air} + r$, then a change in any factor that tends to increase the supply, e.g. to decrease δx, will result in an increase in the concentration around the centre and an increase in the rate. In the limit when the distance between the supply and the reaction centre is very great, the concentration around the reaction centre becomes negligible compared with that in the external supply and $P = r$, i.e. the rate of photosynthesis is determined by the supply of carbon dioxide from res-

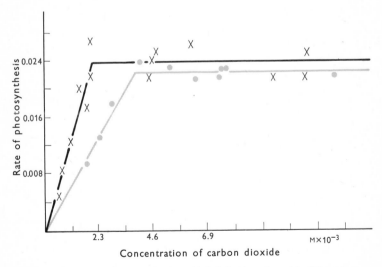

Fig. 1.4   The rate of photosynthesis at high light intensity with different concentrations of carbon dioxide for *Fontinalis antipyretica* (●) and *Elodea canadensis* (×). (Data from Blackman and Smith[13]). Ordinate g CO_2/h/137 cm^2 illuminated surface.

piration. This is called the compensation point, at which it will be noted that in general there is a finite concentration of carbon dioxide in the air above the leaf. Only an increase in supply, e.g. by increasing the external concentration of carbon dioxide or by decreasing δx, will result in an increase in rate.

Experimentally, this type of relationship was first observed by Blackman and Smith[13] when they studied the rate of photosynthesis of leaves of water plants (*Elodea* and a moss) supplied with carbon dioxide by diffusion from a slowly moving film of water (Fig. 1.4). The apparent external concentration required to give half the maximal rate of photosynthesis in such a system is of the order of 1000 times greater than that observed for the suspension of *Chlorella* cells. When, however, the considerations given above are applied to this system, a true value for the affinity, which determines the relationship between rate of photosynthesis and the concentration at the chloroplast, can be deduced. It appears that the constant calculated for the water plants is of the same order of magnitude as that for *Chlorella*. From kinetic studies of a considerable range of plants, the experimental evidence is compatible with the view that the same affinity is shown by the carboxylating enzyme in all plants. It therefore seems likely that all plants use the same carboxylation reaction in photosynthesis.

Limiting factors

The early studies of the kinetics of photosynthesis, undertaken with a variety of plants, served to establish the main qualitative relationship between the rate of photosynthesis, the incident light intensity and the external concentration of carbon dioxide. Just as the rate was related to carbon dioxide concentration in the manner described above, the rate of photosynthesis was found to be proportional to light intensity at relatively low intensities, but to become increasingly independent of intensity at higher light intensities. There was a critical value of light intensity above which the rate of photosynthesis approached independence of light intensity. In 1905 F. F. Blackman deduced from this evidence that photosynthesis involved two reaction steps, a photochemical reaction followed by a dark reaction. He considered that the rate of photosynthesis at lower light intensities was determined by the rate of the photochemical step and referred to this stage as 'light limited'. At higher intensities the rate was regarded as exclusively determined by the characteristics of the dark reaction and referred to as 'light saturated'. Such a hypothesis offered an explanation of the observation that increase of temperature at higher light intensities increased the rate of photosynthesis, but at lower light intensities had little effect (Fig. 1.5). This followed since it is unlikely that a change of temperature will have an appreciable effect on the rate of a

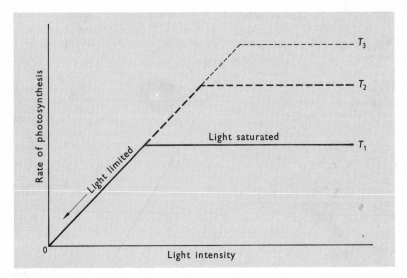

Fig. 1.5 Diagram to illustrate F. F. Blackman's 'principle of limiting factors'.
T_1, T_2 and T_3 are progressively higher temperatures.

purely photochemical process. Similarly the effect of certain poisons, e.g.
cyanide, varied according to whether the effect was measured at low concentrations of carbon dioxide, i.e. CO_2 limited, when the inhibition was
maximal, or observed at high concentrations of carbon dioxide, i.e. CO_2
saturated, when the inhibition was minimal. From such observations O.
Warburg[141] concluded that cyanide primarily inhibited the dark reaction
of photosynthesis, which he believed controlled the overall rate at low
concentrations of carbon dioxide. After Warburg's work, the dark reaction
was referred to as the Blackman reaction. Subsequently it was shown that
photosynthesis is a complex sequence of reactions involving at least two
photochemical processes and many thermochemical processes. Thus
looked at in modern terms the early analysis of the mechanism based on
physiological studies was far too simple to provide an adequate quantitative model of photosynthesis; yet it formed a useful guiding hypothesis for
the study of the influence of internal factors on the rate of photosynthesis
and hence of a major component influencing plant growth.

2

Carbon Metabolism

Introduction

Radioisotopes of carbon have been extensively used in the study of the metabolism of carbon in photosynthesis since the discovery of the isotope of carbon, ^{14}C (half life 5000 years), during the second world war. Before that, attempts were made to study photosynthesis using the isotope ^{11}C, with a half life of only 22 minutes, but the experiments were inconclusive. The stable isotope of carbon ^{13}C is not radioactive and must be used with the mass spectrometer; its use in the study of photosynthesis has been confined to a few kinetic studies.

Ruben, Hassid and Kamen[116] were the first to use ^{11}C in 1939 and the most they could achieve with an isotope of such short life was to analyse the general type of compounds in which radioactivity appeared. They were able to show that almost all the radioactivity appeared during short term photosynthesis in the carboxyl group of an organic acid. They were not able to characterize the compound further.

It was after the war that organic chemists at Berkeley, under the imaginative leadership of Calvin, used the isotope ^{14}C to study the photosynthetic system in detail. They were aided by the recent discovery of the technique of paper chromatography. In the first experiments they added the isotope to photosynthesizing leaves or algae and analysed the compounds which became radioactive.[12] The results of their preliminary experiments with *Chlorella* and with barley leaves are shown in Table 2.1. With short periods of photosynthesis a number of compounds became radioactive. These were certain amino acids, e.g. alanine, aspartate, glutamate, glycine and serine; various carboxylic acids, e.g. succinate, fumarate, glycollate, isocitrate, malate; and sucrose and phosphate esters of sugars. Within a

Table 2.1   Percentage distribution of radioactivity among aqueous alcohol-soluble non-lipid products. (After Benson and Calvin.[12])

Compound	Conditions and period of exposure to isotope						
	30 s PS C	30 s PS B	45 m D C	50 m D B	60 s PS B	5 m PS C	5 m PS B
Alanine	6.1	1.1	11	2.2	7.7	1.9	8.1
Aspartic	0.8	0.1	8.7	28	2.6	3.0	0.7
Glutamic	—	—	23	17	—	0.4	0.8
Glutamine	—	—	1.5	8.2	—	—	—
Glycine	—	12	—	—	3.3	—	4.5
Serine	1.3	4.9	4.0	4.7	3.7	2.0	4 3
Succinic	—	—	17	—	—	0.2	—
Fumaric	—	—	7.5	—	—	—	—
Glycollic	0.5	8.3	—	—	2.0	—	6.4
Isocitric (citric?)	—	—	1.9	6	—	0.1	0.2
Malic	0.9	0.5	21	31	3.3	1.9	2.6
Sucrose	0.5	15	1.2	—	27	23	65
Phosphate esters	89	59	1.2	2.9	50	65	6
Alcoholic insoluble compounds	<5	<5	<5	<5	3.0	45	4.5

Percentages were determined by counting of areas on the paper chromatogram defined by the radiogram. s, seconds; m, minutes; PS, photosynthesis with $C^{14}O_2$; D, dark; C, *Chlorella pyrenoidosa*; B, barley seedling leaves.

quite short time, e.g. 5 minutes, radioactivity was widely distributed throughout most compounds in the cell. There was some indication that barley leaves and *Chlorella* were not quite the same but it was decided, partly for the reasons mentioned in the first chapter and partly for ease of experimentation, that *Chlorella* should be the main object for detailed study. Most of the subsequent measurements were made with *Chlorella* grown under standard conditions in the laboratory with abundant (perhaps super-abundant) carbon dioxide (5% in air) and normally measured with the same concentration.

During measurement the algal suspension was placed in a so-called 'lollipop', a vessel made of two sheets of perspex pressed together; radio-carbon was injected into the suspension and samples taken after measured times of illumination. The algae were killed by running the suspension into boiling alcohol. An alternative way of killing by the use of low temperatures

was also used from time to time but, with hot extraction, the alcohol de-
natured the enzymes, extracted the soluble components from the cells and
stopped the reaction relatively quickly.

A simple time course experiment showed that even in 30 seconds many
different compounds became radioactive (Table 2.1). Therefore it was
necessary to shorten the duration of the experiment in an attempt to limit
the spread of radioactivity to fewer compounds. If the time was short
enough it was hoped that the carbon would have entered only the first
compound and in this way the first stable compound formed in the reaction
could be identified. The time of exposure to radiocarbon was reduced to 5
seconds but even so in *Chlorella* six compounds still became radioactive,
namely alanine, malate, sucrose, two sugars containing 5 or 7 carbon atoms
in the phosphorylated form, and phosphoglyceric acid (PGA).

The discovery of the photosynthetic reduction cycle

Calvin then argued that in a reaction sequence of the type

$$a + CO_2 \longrightarrow b \longrightarrow c \longrightarrow d$$

when the amount of label in b, as a percentage of the total label incorporated,
is plotted as a function of time, in the very shortest periods radioactivity
will be confined to b and nothing else; therefore the nearer we can get to
zero time, the more and more the percentage of radioactivity in b should
approach 100%. Conversely, in a very short period, c will have 0%, because
when b has 100% clearly c must have nothing. The percentage for c begins
to rise as that for b is falling. If the final products are represented by d,
then ultimately all the radioactivity will flow through and end up in d for a
given dose of radiocarbon, and d will ultimately rise to 100% of all the
radioactivity incorporated. The experiment should be such that the tracer
is introduced during steady state photosynthesis; in that case the concentration of intermediates will remain unchanged and only the radioactivity
changes.

In practice there is a limit to the speed at which the experiment can be
completed. After photosynthesis has proceeded for 1.7 seconds the compound with the highest observed percentage is PGA with 70% (Fig. 2.1).
It would be conceivable, however, that within this time another compound
had fallen from 100% and PGA risen from zero. Thus, whilst the time
course of incorporation of radioactivity in *Chlorella* cells gives an indication
that PGA is the compound first formed from carbon dioxide in photosynthesis, it is not a complete proof.

A second line of evidence comes from a study of the sequence of labelling
of the carbon atoms within the molecule. By suitable chemical techniques,

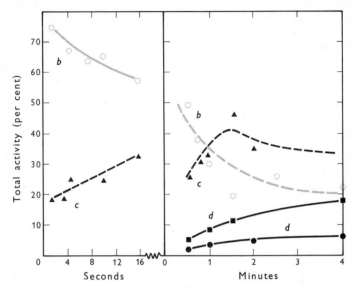

Fig. 2.1 The percentage radioactivity incorporated in different compounds plotted against time, during photosynthesis in the alga *Scenedesmus* in the presence of $^{14}CO_2$. ○, Phosphoglyceric acid; ▲, sugar phosphates; ■, malic and aspartic acids; ●, alanine. *b*, *c* and *d* are the products of the reaction $a + CO_2 \rightarrow b \rightarrow c \rightarrow d$. (After Calvin *et al.* (1951). *Symp. Soc. exp. Biol.*, 5, and Bassham *et al.* (1954). *J. Am. chem. Soc.*, **76**, 1760.)

the three carbon atoms of PGA can be separated as three stable compounds each derived from only one carbon atom of the molecule. With experiments of short duration, i.e between 1 and 5 seconds, all the radioactivity of the molecule was found to come from the carbon atom of the carboxyl group, i.e. CH_2OP—$CHOH$—$C^\star OOH$. This is consistent with its formation by carboxylation of some two-carbon precursor. Calvin and his colleagues spent five years looking for such a precursor substance without success.

Since the essential nature of photosynthesis is the formation of a sugar containing an aldehyde group rather than an acid containing a carboxyl group, there must be a reduction reaction in the process to produce the corresponding aldehyde, glyceraldehyde phosphate (or triose phosphate) from phosphoglyceric acid. Light provides this reducing power since if immediately after illumination cells are exposed to $C^{14}O_2$ in the dark radioactivity accumulates in phosphoglyceric acid without any formation of sugars. If the triose phosphate is derived by direct reduction, it would follow that the carbon atom of the aldehyde group

$$
\begin{array}{cc}
CH_2OP & CH_2OP \\
| & | \\
CHOH & CHOH \\
| & | \\
C{*}OOH & C{*}HO \\
PGA & \text{Triose P}
\end{array}
$$

would contain all the radioactivity of the molecule. Then, by reversing the reactions of the Embden-Meyerhof-Parnas pathway, two molecules of triose phosphate can form one molecule of fructose-1,6-diphosphate by joining together end to end. Thus the carboxyl atom of PGA will provide both the 3 and 4C atoms of the hexose molecule. Therefore radioactivity would be confined to the 3 and 4 carbon atoms of hexose. When the hexose or sucrose sugar formed in a short five-second period of photosynthesis was degraded, it was indeed found that all the radioactivity was in the 3 and 4 carbon atoms.

If the reaction is allowed to continue for 30 seconds or 1 minute the pattern of labelling within the molecules begins to change. Radioactivity now appears in the other two carbon atoms of phosphoglyceric acid and consequently also in the 1 and 6 and 2 and 5C atoms of the hexose.

$$
\begin{array}{ccc}
C{*}HO & C{*}H_2OH & CH_2OP \\
| & | & | \\
CHOH \rightleftharpoons & C{=}O & C{=}O \\
| & | & | \\
CH_2OP & CH_2OP & C{*}HOH \\
 & & | \\
\text{Triose P} & \text{Phospho-diOH acetone} & C{*}HOH \\
 & & | \\
 & & CHOH \\
 & & | \\
 & & CH_2OP \\
 & & \text{Fructose-1,6-diphosphate}
\end{array}
$$

Hence if PGA is formed by carboxylating a precursor, this itself must become radioactive after 30 seconds. Therefore there must be some transfer back from the products of the reaction to form more precursor molecules. Then it was realized that the precursor need not contain only two carbon atoms; it could just as well contain 5 and be regarded as a two-carbon moiety joined to a three-carbon moiety. If on carboxylation the product

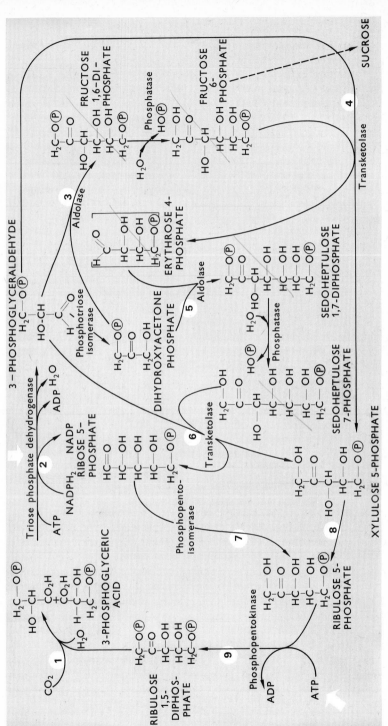

Fig. 2.2 Photosynthetic carbon reduction cycle (Calvin cycle). ↑, Drive from light reaction.

splits, the products would be one three-carbon molecule with a carbon coming from CO_2 and another three-carbon compound (unlabelled). If both molecules were PGA, the fact that two molecules were formed from one precursor molecule would not be detected except that there would on average be only half the activity in the PGA. Ribulose-1,5-diphosphate (RuDP) could be such a precursor. A reaction mechanism for forming a five-carbon sugar from hexose or triose, thus permitting regeneration of the precursor, was postulated by Calvin. About the same time the enzyme transketolase which Calvin postulated as necessary to catalyse this sequence was discovered by Ochoa, Racker and others[67] in animal tissue. A hexose and a triose are condensed together to form a pentose and erythrose. Erythrose combines with another triose to give sedoheptulose, and then sedoheptulose combines with another triose to give two pentoses all in the form of phosphate esters (Fig. 2.2). Thus in total three pentoses are formed from one hexose and three trioses. The pentoses are all converted to ribose monophosphate and then react with adenosine triphosphate to form eventually ribulose diphosphate. This is the precursor molecule which can be carboxylated to form PGA in a reaction catalysed by the enzyme carboxydismutase or ribulose-1,5-diphosphate carboxylase. If this mechanism is correct the sedoheptulose should be equally labelled in the 3, 4 and 5 carbon atoms. Two of the molecules of ribose phosphate should be labelled in C atom 3 only, and the third in C atoms 1, 2 and 3. The distribution of radioactivity observed within these molecules is consistent with this prediction. Thus Calvin was led to postulate a reaction cycle, often called the photosynthetic carbon reduction cycle, since it includes the important step in which PGA is reduced to triose phosphate. The study of the distribution of radioactivity within the intermediate molecules provided important evidence as to the reactions involved in recycling some of the product to form the carbon dioxide acceptor.

The third line of evidence came from the study of the consequences of a sudden change in the conditions of reaction. According to the mechanism just discussed, carbon dioxide enters the cycle at only one point, namely in the carboxylation of ribulose-5-diphosphate to form two molecules of PGA. Therefore when the concentration of carbon dioxide is suddenly reduced, the reaction sequence is slowed at that point and ribulose diphosphate will tend to increase and PGA to decrease in concentration. Thus if in a steady state of photosynthesis the concentration of radioactive CO_2 is rapidly reduced it is to be expected that the radioactivity of RuDP will rise and that of PGA and triose phosphate will fall. Figure 2.3 shows the effect in *Chlorella* when the concentration of carbon dioxide is rapidly decreased from 4% to 0.03%. Within 40 seconds the radioactivity of PGA decreases whilst that in RuDP rises; over longer periods of time more complex changes take place.

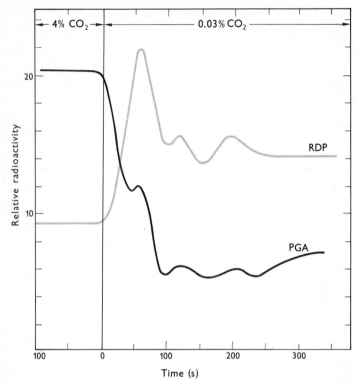

Fig. 2.3 Changes in the concentration of phosphoglyceric acid (PGA) and ribose diphosphate (RDP), following a reduction of carbon dioxide concentration. (After Wilson and Calvin (1955). *J. Am. chem. Soc.*, **77**, 5948.)

A similar interruption of the photosynthetic cycle will occur if the rate of reduction of PGA is rapidly slowed down. Since light energy ultimately supplies the reducing power, if the light intensity is suddenly reduced it would be expected that radioactivity in the sugar phosphates should fall and that in PGA rise. Again this has been observed experimentally during the short period of time immediately following the change.

The reactions of the cycle

The total reaction sequence of the Calvin cycle may be summarized in the equation:

$$3CO_2 + 9ATP + 6NADPH + 6H^+ + 5H_2O \longrightarrow$$
$$1 \text{ triose } P + 9ADP + 8P_i + 6NADP^+.$$

For each molecule of carbon dioxide converted to sugar two molecules of reduced pyridine nucleotide and 3 molecules of ATP are required. There are four important processes in the Calvin cycle. The first is the reduction of phosphoglyceric acid to form phosphoglyceraldehyde

$$PGA + ATP + NADPH + H^+ \longrightarrow triose\ P + ADP + P_i + NADP^+$$

and this is essentially the reverse of the oxidation reaction which takes place in respiration. The reaction is catalysed by triose phosphate dehydrogenase. The enzyme in mitochondria, where the oxidation reactions of respiration take place, is specific for the coenzyme NAD, but there is good evidence that in the chloroplast where the photosynthetic cycle exists the coenzyme is NADP. The oxidation reaction is coupled in respiration to the formation of ATP from inorganic phosphate and to reverse the reaction in photosynthesis a supply of ATP is necessary. We shall see later that light energy supplies both reduced NADP and ATP and together these drive the reaction in the direction of reduction of PGA.

The second process is concerned with the conversion of 5 molecules of triose phosphate into 3 molecules of pentose phosphate. This is essentially

fructose 6P + triose P $\longrightarrow$ erythrose 4P + xylulose 5P
erythrose 4P + triose P $\longrightarrow$ sedoheptulose-1, 7-di P
sedoheptulose-1, 7-di P + triose P $\longrightarrow$ ribose 5P + xylulose 5P
2 xylulose 5P $\longrightarrow$ 2 ribose 5P

a sequence of sugar interconversions.

Thirdly, there is the conversion of the pentose monophosphate to ribulose-1,5-diphosphate, the substrate for carboxylation. Ribulose-1,5-diphosphate is a high energy compound compared with ribose monophosphate (RMP) and ATP is required for the phosphorylation. This permits the free energy of the carboxylation to be negative, and of the order of -28 kJ/mol. Since for the complete reduction of one molecule of CO_2, 2 moles of $NADPH_2$ and 3 of ATP (two for the reduction reaction and one for the phosphorylation of RMP) are required, if the free energy change of hydrolysis of ATP is approximately 32 to 40 kJ/mol and the free energy of oxidation of 2 moles NADP 418 kJ, then approximately 510–540 kJ is available for the sequence of reactions. The free energy of the products relative to the reagents is 470 kJ/mol under standard conditions. Therefore the cycle has a thermodynamic efficiency of between 90–100%.

The fourth reaction, the carboxylation reaction, is still not fully understood. After a transition from light to dark, there is a conversion of RuDP to PGA. From studies of the time course, the rate of flow of radioactivity from RuDP to PGA can be calculated. This shows that in the living cell approximately 1.4 molecules of PGA are formed for every molecule of RuDP used.[9] Again *in vivo* when the concentration of CO_2 is increased to a higher level, RuDP is carboxylated to form PGA. When the specific activity lost from the RuDP is compared with that of the PGA formed, the radioactivity in the PGA is consistent with the formation of less than two molecules of PGA from one of RuDP. But if the proteins of the chloroplast are fractionated, partially purified fraction 1 protein (i.e. the larger portion of the soluble protein of mol. wt. 18–21 S) catalyses the reaction between ribulose-1,5-diphosphate and CO_2 *in vitro* and produces two molecules of PGA for every mole of RuDP consumed.[113]

$$CO_2 + CH_2OP - \overset{\overset{\displaystyle O}{\|}}{C} - CHOH - CHOH - CH_2OP + H_2O$$
$$\rightarrow 2(CH_2OP - CHOH - COOH)$$

Hence in the plant it appears that RuDP does not give rise to the equivalent amount of PGA and must in part give rise to something else. Furthermore, *in vitro* the reaction is only rapid in the presence of relatively high concentrations of carbon dioxide, i.e. 0.1%, indicating a k_m for the isolated enzyme of 4.5×10^{-4}M free CO_2, whereas intact plants grow well in air with only 0.03% CO_2. The activity of the enzyme *in vitro* is activated by preincubation with magnesium ions and with the acceptor ribulose-1,5-diphosphate.[71] This may occur to some extent *in vivo* due to ionic movements induced in chloroplasts by illumination (see Chapter 8). Even so, the present view is that the enzyme has been partially inactivated during extraction. In the plant the enzyme may be part of a large organized protein complex to which ribulose-1,5-diphosphate is bonded with a particular spatial orientation. The bond at one end of the molecule probably involves a SH linkage and a metal group (probably copper), since the reaction is inhibited by iodoacetamide and cyanide. Wishnick and Lane[146] showed that cyanide binds to the enzyme stoichiometrically but only in the presence of the substrate RuDP. Carbon dioxide enters the middle of the RuDP molecule becoming affixed to carbon atom 2, the complex splitting[105] between carbon atoms C2 and C3 of the original molecule, thus forming one molecule of PGA which contains the newly introduced CO_2. During photosynthesis (in the presence of excess reduced ferredoxin) one half of the molecule may be reduced *in situ*; when the enzyme releases this fraction it is as triose phosphate and PGA is not released in a free form. Hence *in vivo* the carboxylation and reduction are co-ordinated, some reduction of PGA occurring on the carboxylating enzyme site as well as after release in the free form. Hence the kinetics of the *in vitro* carboxylation differ from those observed *in vivo*.[22]

Criticisms of the cycle

When a more detailed investigation of the distribution of radioactivity within the molecules formed in the early seconds of photosynthesis was undertaken a discrepancy was found. Originally Calvin and Bassham had claimed that after fixation for a period of 15 seconds, there was equality in the 1 and 2C atoms of phosphoglyceric acid, and furthermore that in the hexose formed from it there was not only equality between the 3 and 4, but also between the 1 and 6 and 2 and 5C atoms. Kandler and Gibbs[76] assayed all six carbon atoms independently and found the 4th C atom was always labelled to a greater extent than the 3rd and the 1st and the 2nd to a greater extent than the 5th and 6th. It will be noted that this cannot be simply attributed to the union of a more active and a less active 3 C triose since it is the 4 C atom which is labelled greater than the 3, yet the 5 and 6 are labelled less than 1 and 2.

With respect to the discrepancy between C atoms 3 and 4, Bassham has noted that C atoms 1, 2 and 3 of hexose are derived from dihydroxyacetone phosphate and 4, 5 and 6 from phosphoglyceraldehyde. Hence if there were a larger pool of dihydroxyacetone phosphate than of triose P in photosynthetic cells, equilibration with this large unlabelled pool would dilute the radioactivity and result in one half of the hexose molecules being less radioactive than the other half. Gibbs[51] and other workers subsequently argued that if this were correct and they studied isolated chloroplasts rather than whole cells, since the pool size would be expected to be smaller, the dilution effect should be smaller. But they obtained essentially the same result with chloroplasts isolated from spinach leaves as with whole algal cells. Trebst and Fiedler[134] extended these observations by using broken chloroplasts which were fed either radioactive carbon dioxide or phosphoglyceric acid labelled in the carboxyl group. When carbon dioxide was fed they again observed that carbon atom 4 was labelled to a greater extent than 3 in fructose-1,6-diphosphate, but when labelled phosphoglyceric acid was used, the hexose formed was equally labelled in both the 3 and 4 carbon atoms. It followed that the discrepancy between C atoms 3 and 4 could not arise from any reactions occurring between phosphoglyceric acid and hexose. The discrepancy must arise in the reactions related to the formation of PGA from carbon dioxide and they suggested that this would be consistent with the view that the carboxylation *in vivo* gives rise to one molecule of phosphoglyceric acid released immediately from the enzyme complex, and a second molecule released possibly only after reduction on the enzyme site. This can preferentially give rise to two pools of triose P which will differ in their average level of radioactivity.

We have now to consider the other discrepancy, namely why carbon atoms 1 and 2 are labelled more than 5 and 6. Bassham[9] has suggested that this may be due to the fact that the 1 and 2 carbon atoms of both

hexose and sedoheptulose are concerned in transfers of 'active glycoaldehyde' catalysed by transaldolase and transketolase in the photosynthetic
cycle. There may be some equilibration between the two-carbon moiety
transferred in these reactions and the 1 and 2 C atoms of the hexose. The
reactions involving 2 carbon transfers are readily reversible since the free
energy of the reaction is only of the order of 4 kJ/mol. Since the 2-carbon
fragment is also in equilibrium with the pentose pool, the radioactivity in
the 1 and 2 carbon atoms of ribose-5-phosphate tends to equilibrate
through the 2-carbon moiety with the 1 and 2 atoms of fructose-1,6-
diphosphate. In all cases the magnitude of the effects will depend to a large
extent on pool sizes and will vary according to the conditions.

It has also been suggested that the carboxylation reaction *in vivo* may
give rise to an unstable short-lived intermediate which has escaped detec-
tion in the chemical analytical procedures used because it breaks down
readily to form PGA. When the time course of total carbon incorporated
by a cell suspension was compared with the arithmetical sum of the carbon
incorporated into all known intermediates and extrapolated back to short
times, a discrepancy of 1.5 μM C/mm^3 packed *Chlorella* cells was observed.
Since the rate of photosynthesis is 12 μmoles carbon/min/mm^3 of cell, this
discrepancy approximates to the equivalent of fixation for 7 seconds. A
number of investigators have attempted to find a compound to account for
this discrepancy by the use of reagents forming addition compounds, e.g.
cyanide, or by the use of low temperatures for killing, extraction, and for
the chromatographic analysis. However, extensive investigations, particularly by Rohr and Bassham,[115] gave little evidence of the existence of such
compounds in either *Chlorella* or the bean leaf. Some discrepancy between
the total carbon incorporated and the sum total incorporation in all the
stable intermediates could be accounted for by the formation of bicarbonate
in the acid sap of the cells. If it is assumed that the pH of the cell is near 7,
then the carbon dioxide equivalent of the bicarbonate present is of the
order of 1–1.5 μmoles per cell and this could largely account for the discrepancy between the total fixation and the sum total of all intermediates.
It seems unlikely that any significant intermediates of the photosynthetic
cycle remain undiscovered.

There have also been suggestions, notably by Stiller,[124] that carbon may
be introduced into intermediates of the photosynthetic carbon cycle, by
means of a second carboxylation in addition to the carboxylation of
RuDP. If this is the case, then from a knowledge of the pool sizes of each
of the intermediates contained in the relevant cycles, we can predict the
speeds at which the pools of these intermediates should incorporate label.
For example, again in *Chlorella*, the pool size of RuDP is 0.36 μmoles, that
of PGA 3 μmoles and that of triose phosphate 0.8 μmoles (all per mm^3
packed cell volume). Introduction of carbon dioxide into the reaction cycle

should result in an increase with time of the specific activity of the phosphoglyceric acid at a certain rate. If however carbon dioxide were also introduced independently, say through a diose and the concentration of diose
were of the order of $1.5 \ \mu mol/mm^3$ cells, the rise in specific activity of the
PGA would now be slower than in the former case. Bassham has shown
that the observed value is in better agreement with the calculated value
when no secondary introduction of carbon dioxide and no unidentified
intermediate of the cycle is assumed.

There is increasing evidence that, in certain plants, a considerable incorporation of radioactivity occurs into other compounds prior to incorporation into phosphoglyceric acid. This was first observed by Kortschak, Hartt
and Burr[85] in sugar cane plants, and was subsequently confirmed by Hatch
and Slack.[55] In such plants the initial fixation is largely into malic and
aspartic acids, e.g. in one second as much as 93% of the carbon incorporated can be accounted for by these two four-carbon acids. Radioactivity
appeared in phosphoglyceric acid only after relatively longer periods of
photosynthesis. Furthermore, the time course observed was not greatly
affected by changes in the partial pressure of carbon dioxide or of light
intensity. The radioactivity was largely incorporated initially into carbon
atom 4 of malate and later in carbon 1 of PGA. If after treatment with radioactive carbon dioxide, plants were transferred to normal carbon dioxide,
although the total radioactivity remained almost unchanged, a large decrease in the radioactivity present in malate and aspartate and a considerable increase in radioactivity in sucrose was observed. Hatch and Slack
therefore suggested that in such plants carboxylation took place preferentially of phosphoenolpyruvate (PEP) to form oxalacetate which was then
converted into malate and aspartate, and that

$$PEP \qquad + CO_2 + H_2O \longrightarrow \qquad Oxalacetate \qquad\qquad + P_i$$

$$COOH.COP{=}CH_2 \qquad\qquad\qquad COOH.\overset{\displaystyle O}{\overset{\|}{C}}{-}CH_2{-}COOH$$

these acids subsequently transferred carbon to sugar phosphates which
then ultimately appeared in sucrose. Hatch and Slack surveyed a number of
plants with respect to the time course of photosynthetic incorporation of
^{14}C, and found that of 32 plant species investigated, 14 behaved in a similar
way to sugar cane; these included various varieties of sugar cane, sorghum,
maize and other tropical grasses. Such plants are often referred to as C^4
plants (as distinct from C^3 or Calvin type plants). They suggested that in
these plants there were two cyclic processes linked together by a carbon
transfer reaction. The role of the second cycle was to fix carbon dioxide
and thus to provide carboxyl groups to react with a carbon acceptor in the

primary cycle. After phosphoenolpyruvic acid was carboxylated to form oxalacetate, the oxalacetate in the form of malic acid was considered to react with ribulose diphosphate, transferring one carbon to the pentose and the remaining three appearing as pyruvate (Fig. 2.4). From this the phos-

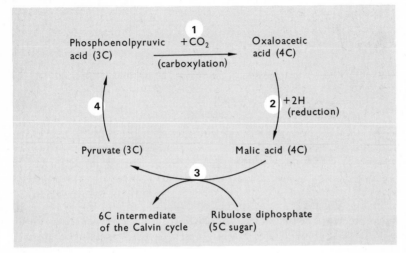

Fig. 2.4 Accessory cycle to the Calvin cycle proposed by Hatch and Slack to operate in some plants. CO_2 is used to produce oxaloacetic acid from phosphoenolpyruvic acid (reaction 1). In reaction 2 oxaloacetic acid is reduced to malic acid which reacts with ribulose diphosphate, forming the 6C intermediate of the Calvin cycle (reaction 3). The remaining 3C fragment of the malic acid appears as pyruvic acid which is converted to phosphoenolpyruvic acid (reaction 4), thus completing the accessory cycle.

phopyruvic acid required for the carboxylation could be reformed by a reaction catalysed by phosphopyruvate synthetase, an enzyme discovered by Hatch and Slack and shown to be confined to higher plants which show a C^4 fixation pattern. The ribulose diphosphate was part of the same primary cycle as Calvin had postulated. Thus instead of incorporation of carbon dioxide direct into the photosynthetic reduction cycle in these plants, carbon dioxide is first added to form a 4C acid and this serves to add carbon to RuDP instead of CO_2; however these reactions require two additional molecules of ATP per CO_2 molecule incorporated.

Plants which exhibit this type of metabolism, C^4 plants, normally have a specialized anatomy in which the vein of the leaf is surrounded by a bundle sheath consisting of a specialized layer of cells. Chloroplasts are most abundant in the cells of the bundle sheath and in the first layer of mesophyll cells immediately adjacent. Starch grains occur predominantly in the bundle sheath cells. It has proved possible to isolate separately the two types of cells. Photosynthesis of the isolated cells shows predominantly a Calvin cycle metabolism in the bundle sheath cells and a C^4 acid metabolism in the mesophyll cells (Fig. 2.5). An important difference claimed

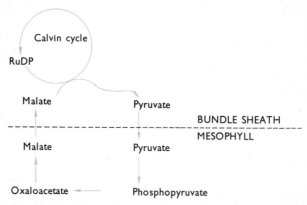

Fig. 2.5 Transfer of malate and pyruvate between mesophyll and bundle sheath cells during photosynthesis according to the hypothesis of Hatch and Slack.

between the two cell types was the relatively greater activity of the enzyme PEP carboxylase in the mesophyll cells and the greater activity of RuDP carboxylase in the bundle sheath cells. However, it is possible that some enzyme activity in the mesophyll cells is inhibited by the liberation of phenolic compounds in the early stages of grinding and that the enzymes which catalyse the Calvin cycle reactions are inhibited to a greater extent by these compounds. Attempts have been made to isolate chloroplasts from the two types of cell and to study their carbon metabolism. Bundle sheath chloroplasts in the main lack grana and have internal lamellae extending throughout the length of the plastid (agranal), whereas the mesophyll chloroplasts have the normal grana type of organization. Of many attempts only Gibbs et al.[52] have succeeded in obtaining chloroplasts isolated from the mesophyll cells of maize which are capable of assimilating CO_2. They found identical products in chloroplasts from both types of cells similar to those of chloroplasts isolated from spinach or pea. The possibility exists that C^4 plants differ from C^3 only in having an abundance of PEP carboxylase in the mesophyll cells and that this enzyme is localized in the cytoplasm. CO_2 entering the cell will be preferentially incorporated by this enzyme (and similarly as we shall see in Chapter 3, any CO_2 produced by the chloroplasts in the process of photorespiration would be fixed by this enzyme and would not escape from the leaf). Thus the additional enzyme forms a reservoir of fixed CO_2 in the cytoplasm of the mesophyll cells. In earlier studies, Hatch and Slack[56] considered that the primary function of the mesophyll cells was to incorporate CO_2 into malic acid and that the malic acid then moved into the bundle sheath cells to donate carbon to RuDP, the pyruvate formed returning to the mesophyll cell where it was again carboxylated. However, the weight of the evidence is now against the movement of malic acid between cells and it is more likely that the function of PEP carboxylase is to accumulate CO_2 in the mesophyll cells.

Another reaction capable of forming malate from pyruvate requires reduced pyridine nucleotide and is catalysed by malic enzyme,

$$Pyruvate \quad + CO_2 + NADPH + H^+ \xrightarrow{[Mg^{2+}]}$$

$$\underset{\displaystyle COOH \cdot C \cdot CH_3}{\overset{\displaystyle O}{\underset{\|}{}}}$$

Malate $+ NADP$

$$COOH \cdot CHOH \cdot CH_2 \cdot COOH$$

The free energy change of this reaction favours decarboxylation rather than carboxylation at levels of carbon dioxide near that in normal air $[k_m = 4 \times 10^{-3} M]$. The same is true for the conversion of ribulose 5P to 6-phosphogluconate catalysed by 6-phosphogluconate dehydrogenase. It is unlikely that either of these reactions using reduced pyridine nucleotide is responsible for significant fixation of carbon dioxide in photosynthesis.

3

Photorespiration

The formation of glycollic acid during photosynthesis

The photosynthetic reduction cycle of Calvin leading to sucrose and other sugars as the predominant product was deduced from studies of photosynthesis on plants exposed to high concentrations of carbon dioxide. In algae, if the carbon dioxide concentration is decreased until the rate of photosynthesis is proportional to its concentration, glycollic acid appears as a product of photosynthesis. The accumulation of free acid rises to an optimum at a relatively low concentration of carbon dioxide;[145] also, in higher plants glycine and serine appear as photosynthetic products. When the substance isonicotinylhydrazide is added to *Chlorella* it is found to prevent the subsequent metabolism of glycollate, the accumulation of glycollate is greatly increased and the optimum with respect to concentration of carbon dioxide much emphasized (Fig. 3.1); it does not alter significantly the rate of photosynthesis. Furthermore, the higher the light intensity, the more glycollate is accumulated, and indeed the production of glycollate continues to rise even for light intensities above those when photosynthesis is approaching 'saturation'. In the presence of isonicotinylhydrazide and at the highest light intensities approximately 80–90% of the carbon incorporated may accumulate as glycollate. If isonicotinylhydrazide is doing nothing more than preventing the subsequent metabolism of two carbon compounds, it follows that in the absence of inhibitor there must be a flow through these compounds of the order of 80 or 90% of the total carbon incorporated. Similar results have been observed if instead of isonicotinylhydrazide, sodium hydroxy-pyridine methylsulphonate is added to photosynthesizing tobacco leaves; with excised tobacco leaves Zelitch[149] observed a 20-fold accumulation of glycollate within 30 minutes.

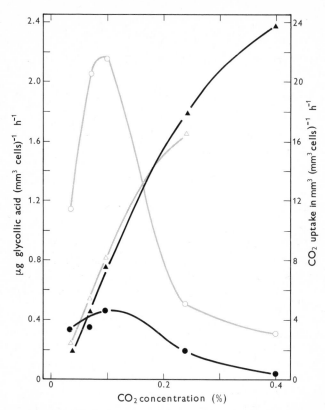

Fig. 3.1   The effect of carbon dioxide concentration on glycollic acid excretion and the rate of carbon dioxide uptake in the presence and absence of isoniazid in the alga *Chlorella*. In the absence of isoniazid: ●, glycollic acid excretion; ▲, carbon dioxide uptake. In the presence of isoniazid: ○, glycollic acid excretion; △, carbon dioxide uptake. (After Whittingham and Pritchard.[145])

This substance of the general formula $R—CHOH—SO_3Na$ inhibits the enzyme glycollic oxidase since it acts as a competitive inhibitor with the substrate $CH_2OH—COOH$. As mentioned in Chapter 1 this substance also causes stomatal closure but the relationship if any between this effect and the effect on glycollate metabolism is not yet clear. If all these facts are to be consistent with the hypothesis that the Calvin cycle has only one reaction for the uptake of CO_2, the sequence of reactions must be modified so as to allow under certain conditions almost all the carbon entering to appear as glycollate. The conditions for maximal glycollate formation are that the CO_2 concentration shall be relatively low, i.e. the carboxylation reaction is slowed down to a maximum extent, and the light intensity be high, i.e. the reduction step is maximal, permitting maximal formation of acceptor for carbon dioxide. Thus the formation of two carbon compounds is greatest when the accumulation of RuDP might be expected to be maximal. If we regard RuDP as made up of two moieties of two and three C atoms respectively, with a potentiality for cleavage, the 2C moiety might separate as a free entity to form phosphoglycoaldehyde,

$$
\begin{array}{lll}
\text{CH}_2\text{OP} & \text{CH}_2\text{OP} & \\
| & | & \text{P-glycoaldehyde} \\
\text{C}{=}\text{O} & \text{CHO} & \\
\underline{\quad\quad}|\underline{\quad\quad} & & \\
\text{CHOH} \longrightarrow & \text{CHO} & \\
| & | & \\
\text{CHOH} & \text{CHOH} & \text{Triose P} \\
| & | & \\
\text{CH}_2\text{OP} & \text{CH}_2\text{OP} &
\end{array}
$$

the remaining moiety being the same three carbon compound normally formed by carboxylation and reduction, i.e. phosphoglyceraldehyde. So far, attempts to demonstrate this cleavage reaction *in vitro* have not been very successful.[14] Removal of the phosphate groups from phosphoglycoaldehyde and subsequent oxidation would produce free glycollic acid ($CH_2OH \cdot COOH$).

Glycollic acid can be oxidized to glyoxylate and then aminated to form glycine ($CH_2NH_2 \cdot COOH$), and from two glycine molecules, with elimination of CO_2, serine can be formed ($CH_2OH \cdot CHNH_2 \cdot COOH$). Under normal conditions the serine could then transaminate with oxaloacetic acid to produce aspartic acid or with α-ketoglutarate to form glutamate, thus contributing to the general nitrogen metabolism of the cell. Analysis of the compounds formed from radioactive carbon dioxide in the presence of isonicotinylhydrazide show that the formation of serine from glycine is inhibited and consequently there is an accumulation of glycollate and glycine.

If, in *Chlorella*, photosynthesis in the presence of $^{14}CO_2$ continues until all the intermediates are radioactive and the gas supply is then rapidly changed from $^{14}CO_2$ to $^{12}CO_2$, the radioactivity in phosphoglyceric acid immediately begins to be diluted as $^{12}CO_2$ is taken up[145] (see Fig. 3.2). Progressively, radioactivity is lost from the sugar diphosphates and monophosphates and within 20 minutes the radioactivity of compounds in the photosynthetic cycle has fallen almost to zero. The addition of isonicotinylhydrazide does not alter the rate of loss of radioactivity from the Calvin cycle intermediates. (Allowance must be made for the fact that in the

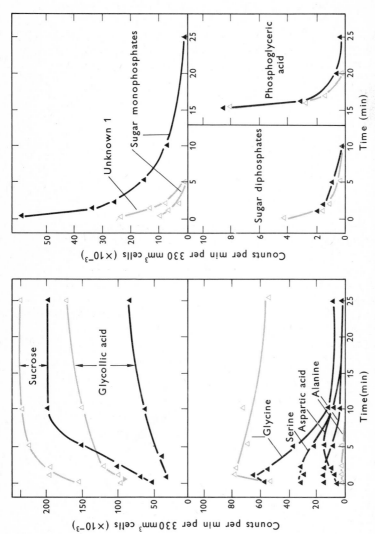

Fig. 3.2 The time course of loss of radioactivity in individual compounds when ^{14}C is replaced by ^{12}C in the gas phase during photosynthesis with *Chlorella pyrenoidosa*. ▲, In the absence of INH; △, in the presence of INH. (After Whittingham and Pritchard.[145])

presence of the inhibitor certain sugar phosphates form a complex and new spots appear on the chromatogram.) As the intermediates of the cycle lose radioactivity, the radioactivity begins to rise in glycollic acid, and this occurs to a greater extent in the presence of isonicotinylhydrazide. Sucrose is another compound which begins to accumulate radioactivity and this also rises to a greater extent in the presence of isonicotinylhydrazide. The proportion of sucrose to glycollate is related to the concentration of carbon dioxide present. Subsequently the radioactivity in glycine begins to increase but then decreases again as other amino acids become radioactive, e.g. serine, aspartate and alanine. However, the radioactivity does not disappear from glycollate. Since glycollic acid is a relatively small molecule, which can permeate the cell wall of *Chlorella*, part of it escapes into the external medium when it accumulates as free glycollate.[133] Clearly that part of it which has escaped from the cell cannot be further metabolized. In the presence of isonicotinylhydrazide the radioactivity entering glycine does not subsequently fall but stays relatively constant with time and virtually no activity appears in serine. This suggests that isonicotinylhydrazide primarily inhibits the metabolism of glycine to serine.

The formation of two carbon compounds is a consequence of driving the photosynthetic cycle under conditions where the carboxylation of RuDP is reduced relative to its rate of formation. If glycollate is derived from carbon atoms 1 and 2 of ribulose or of other pentose molecules it should be equally labelled in its two C atoms. If, on the other hand, it arises from an independent carboxylation there are two possibilities: either CO_2 would be added to a 1 carbon acceptor molecule to form glycollate labelled predominantly in only one carbon, or two molecules of carbon dioxide might condense as Stiller[124] proposed to form an equally labelled diose. The experimental evidence suggests that both C atoms of glycollate are equally labelled for short periods of photosynthesis. It follows that the two C atoms of glycine should also be equally labelled and serine labelled equally in all three C atoms. Serine might also be formed by an alternative route from phosphoglyceric acid after conversion to hydroxypyruvate, phosphoserine and hence serine.

$$CH_2OP \cdot CHOH \cdot COOH \longrightarrow CH_2OP \cdot CO \cdot COOH \longrightarrow$$
$$\text{Phosphoglycerate} \qquad\qquad \text{Phosphohydroxypyruvate}$$

$$CH_2OP \cdot CHNH_2 \cdot COOH \longrightarrow$$
$$\text{Phosphoserine}$$

$$CH_2OH \cdot CHNH_2 \cdot COOH$$
$$\text{Serine}$$

Serine which originated from these reactions would have the same labelling as was characteristic of PGA; hence for short periods of photosynthesis when labelling of the carboxyl of the PGA is predominant the serine formed in this manner would be unequally labelled. Experimentally it is found that the distribution of radioactivity in serine varies according to the conditions; under conditions where the flow through two carbon compounds is maximal the serine produced is equally labelled in all three carbon atoms. This is consistent with the formation of this acid from glycollate and glycine.

Glycollate also appears as a major product of photosynthesis when the oxygen concentration is high relative to the concentration of carbon dioxide.[26] If the oxygen concentration is increased up to 50 or even approaching 80% then again 80% of the carbon incorporated in photosynthesis by *Chlorella* appears as glycollate. The influence of oxygen partial pressure is most apparent at low concentrations of CO_2, suggesting that oxygen probably stimulates some of the reactions which are dominant under conditions of slow carboxylation, i.e. the cleavage of RuDP, resulting in the formation and metabolism of two carbon compounds. Oxygen may stimulate the oxidation of glycoaldehyde to glycollate, so that raising the partial pressure of oxygen will result in an increase of glycollate formation ultimately at the expense of the other product of the Calvin cycle, i.e. sucrose. In *Chlorella*, studies of the time course of labelling with $^{14}CO_2$ show relatively little accumulation of glycollate in 20% O_2 but, for the same concentration of CO_2, oxygen concentrations between 40 and 100% produce a large accumulation of glycollate. The effects of isonicotinyl-hydrazide and of oxygen in stimulating the accumulation of glycollate act independently. One important difference is that increased oxygen inhibits the rate of photosynthesis; this is consistent with the proposal that it decreases the pool size of intermediates in the Calvin cycle by stimulating the oxidation of pentose to phosphoglycollic acid.

The compensation point

When a leaf is brightly illuminated in an enclosed space the concentration of CO_2 in the gas phase falls with time. It does not approach zero but only a finite value called the 'compensation point' which is independent of light intensity provided this is moderately high (see Fig. 3.3). This suggests that certain reactions in the light must produce CO_2 and others consume it; when the two processes just balance there is no net change. The process producing CO_2 in the light has been called photorespiration. An alternative procedure is to commence the experiment with no carbon dioxide in the gas phase when upon illumination the concentration rises until it reaches the same compensating value. The concentration that can be measured in such a closed system in the steady state is a measure of the concentration of CO_2 in the intercellular spaces of the leaf when no net

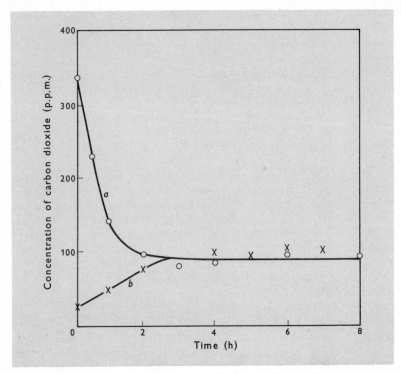

Fig. 3.3 Time course of carbon dioxide concentration changes in a closed vessel of air: *a*, with leaflets of *Sambucus nigra*; as *a* but with an initial concentration of 24 p.p.m. carbon dioxide. Light intensity 10 000 lux in all cases. (After Gabrielsen, E. K. (1948). *Nature, Lond.*, **161**, 138.)

movement of carbon dioxide is taking place between the leaf cells and the external air. The leaves of many different species give approximately the same compensation point between 50 and 100 p.p.m. CO_2 with the exception of a group of tropical grasses and related species which include maize, sugar cane and certain species of *Amaranthus* and *Atriplex* (Table 3.1). The compensation concentration approximately doubles for a 10°C rise in temperature. The latter group of plants have compensation values very near to zero and also show a special pattern of early products of photosynthetic carbon fixation (see Chapter 2). They do not liberate CO_2 into the external air in the light. It does not follow that these plants do not possess the enzyme systems responsible for photorespiration since it could equally be that the CO_2 released by photorespiration is refixed by the leaf cells before it escapes into the external atmosphere. Therefore it cannot be

Table 3.1   Carbon dioxide compensation point of various species in air[103] at 23°C

Plant	p.p.m.
Maize	9
Sugar cane	7
Tobacco	60
Wheat	52
Tomato	75
Norway maple	145

concluded that a higher value for the compensation point necessarily implies a higher rate of photorespiration. In general, the same features of structure and anatomy which influence the uptake of carbon dioxide and its release will influence the compensation point.

Increasing the partial pressure of oxygen in all plants even within the range of 2% to 50% oxygen greatly increases the compensation point.[43] This may be attributed to the effect of oxygen in increasing the rate of photorespiration because dark respiration would be relatively little affected by a change in oxygen partial pressure in this range; presumably the release of carbon dioxide in photorespiration is accompanied by oxygen uptake.

The measurement of photorespiration

Attempts have also been made to determine the rate of photorespiration by the use of isotopes of oxygen which can be used to separate the oxygen evolved in photosynthesis and that consumed in respiration. A. H. Brown[16] was able to measure the rate of oxygen evolution in a green alga in the light from the consumption of ^{18}O from the gas phase; the oxygen evolved in photosynthesis largely came from water and the water present was $H_2{}^{16}O$ (see Fig. 3.4). Using this technique, Brown was able to show that under many conditions the rate of oxygen uptake in the light was little different from the rate of uptake in the dark. Most of his measurements were confined to relatively high concentrations of CO_2 and low concentrations of oxygen when the rate of photorespiration was likely to be small. There has also been criticism that in the design of these experiments only that gas which is not reconsumed by the plant tissue can escape into the gas phase and be measured and in Brown's experiments the analyses were made exclusively on the gas phase (by sampling with a mass spectrometer). In later experiments Hoch and Kok[66] were able to analyse the isotopic composition of the liquid phase in which the photosynthetic algae were sus-

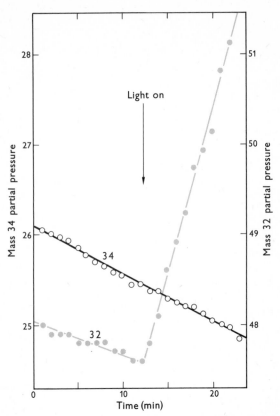

Fig. 3.4   Differentiation between the simultaneous processes of oxygen uptake due to respiration and of oxygen production due to photosynthesis in light, using oxygen isotopes. ^{34}O was present in the gas phase, and its consumption is proportional to the rate of respiration. ^{32}O was present in the water also and is consumed in the dark, but consumed and produced in the light due to 'water splitting' in photosynthesis. Calculation shows that the rate of oxygen consumption is unchanged by illumination. (From van Norman, R. W. and Brown, A. H. (1952). *Pl. Physiol., Lancaster*, **27**, 691.)

pended. Under these conditions they found in *Scenedesmus* that there was little difference in the rate of oxygen uptake in light and dark except at concentrations of CO_2 lower than the compensation point. At that concentration, or lower, there was a large increase in oxygen uptake with increased illumination. For technical reasons it is not possible using the mass spectrometer to make observations with high sensitivity at other than rather low partial pressures of oxygen. Hence the conditions of the experiment are not optimal for the observation of photorespiration.

More convincing evidence of the stimulation by light of the release of CO_2 by green cells has come from studies with radioactive carbon. If after photosynthesis in $^{14}CO_2$ a plant is placed in an atmosphere devoid of all carbon dioxide but kept in the light, photorespiration will continue and liberate $^{14}CO_2$. The rate of evolution is greater the higher the partial pressure of O_2.[53] A similar change in the external concentration of O_2 in the dark does not affect the rate of production of carbon dioxide in normal dark respiration. If after a subsequent period of darkness, during which $^{14}CO_2$ is released at a low rate, the light is put on again, the production of $^{14}CO_2$ shows a marked increase. When the specific activity of the CO_2 produced in the dark and light is determined, the value in the dark is characteristic of the average ratio of radioactive compounds and non-radioactive compounds present in the plant. But the specific activity of the CO_2 produced in the light, i.e. the ratio of radioactive to non-radioactive molecules, is much greater and hence the process of photorespiration must use carbon substrates which have a greater specific activity than the average cell contents. They must be derived only from substances formed most recently in photosynthesis. This might suggest that the process of photorespiration is localized within the chloroplast or near to it.

Before the use of radioactive carbon isotopes, limited experiments were undertaken with the infra-red gas analyser determining total carbon dioxide evolution. In such experiments an excess release of CO_2 by the green leaf was observed in the first seconds of dark immediately following a period of illumination:[29] the short period during which the output of CO_2 exceeds the subsequent dark steady state rate defines the so-called 'CO$_2$ burst'. The magnitude of the burst is increased when the oxygen partial pressure is increased in the range from 2–50% oxygen.[135] The burst is very much smaller in maize and other plants with a low compensation point and it seems probable that the size of the CO_2 burst is correlated with the rate of photorespiration in the preceding light period. Although the response of stomata is probably slow compared with the periods used for such measurements, nevertheless changes in stomatal aperture cannot be totally discounted as having some influence in this type of observation.

The influence of the partial pressure of oxygen on the rate of photosynthesis had been known for many years prior to the use of radioactive carbon dioxide. Using gasometric techniques to compare the rate of photosynthesis at different partial pressures of oxygen and of CO_2, it had been shown that oxygen inhibited the rate of photosynthesis in leaves of many crop plants particularly at low concentrations of carbon dioxide. Most recently Krotkov and his collaborators in Canada[87] investigated this relationship in detail for the soya bean leaf. They found that when the apparent rate of photosynthesis is plotted against the CO_2 partial pressure at different partial pressures of oxygen, the slope of the line is decreased for each increase of oxygen, i.e. the efficiency of the fixation process is reduced (see Fig. 3.5). But also the curves showed that the intercept on the CO_2 axis changed with different oxygen tensions. The intercept is a measure of the minimum concentration of carbon dioxide which must be exceeded to have net photosynthesis, i.e. the carbon dioxide compensation point. This was found to be higher the greater the O_2 tension.

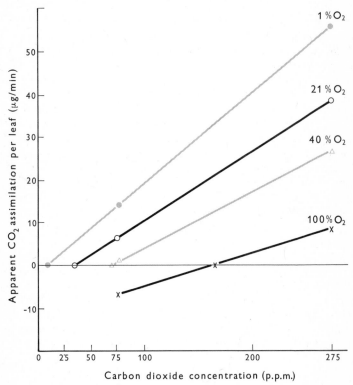

Fig. 3.5   Relation of apparent assimilation of carbon dioxide by soybean leaves to carbon dioxide concentration at different oxygen concentrations. (After Forrester et al.[43])

The reactions which produce CO_2 in the light in green tissue have been called photorespiration and the observations discussed in this chapter suggest that this process must be sensitive to oxygen for a wide range of concentrations. This distinguishes it from the process of dark respiration which for most plants saturates at concentrations less than 1% oxygen. Photorespiration increases with oxygen concentration up to a rate five to ten times the maximum rate of CO_2 production in the dark; it is not saturated even with 50% oxygen.

The substrate of photorespiration

One reaction related to the photosynthetic cycle which results in the evolution of carbon dioxide is the conversion of two molecules of glycine derived from glycollate to form one of serine and one of CO_2. Such a mechanism of CO_2 production would be light dependent, since the metabolism of 2C compounds is a consequence of photosynthetic metabolism and does not take place to a significant extent in the dark. It follows that if 2C compounds are the substrate for photorespiration the rate will be higher the lower the partial pressure of CO_2 and the higher the oxygen tension; when the concentration of CO_2 is high, photorespiration will be at a minimum. Addition of isonicotinylhydrazide or hydroxysulphonate should decrease the production of CO_2 by the mechanism proposed. All these features have been observed to be characteristic of photorespiration and suggest that glycollate metabolism results in CO_2 evolution in the light.

Plants exhibiting the fixation pattern described by Hatch and Slack do not show significant rates of carbon dioxide release in the light. One explanation put forward by Hatch and Slack suggested that such plants avoided the reaction involving the carboxylation of RuDP but used carbon transfer from a 4C acid to RuDP. This hypothesis has not been supported by *in vitro* enzymic studies. This type of plant (the C^4 plants) has been shown to possess the enzymes which catalyse the metabolism of glycollic acid, since feeding exogenous glycollate to maize leaves has been shown to result in sucrose formation. But when maize leaves or tobacco leaf segments are placed in a high concentration of oxygen after exogenous feeding with glycine, they produce much less CO_2 than tobacco. Also if the oxygen partial pressure is increased maize will release CO_2 in the light but the rate is lower even with a relatively high oxygen and a relatively low CO_2 concentration.

A useful manner of distinguishing the two types of plants is by mass spectrographic analysis. Plants with a high rate of release of carbon dioxide in light also prove to have a smaller natural content of ^{13}C in relation to ^{12}C. The difference in parts per 1000 of ^{13}C relative to ^{12}C (taken as a proportion of ^{13}C to ^{12}C in a standard carbon sample) is approximately 113 for maize, sugar cane, sorghum, and so on, but 27 for wheat and barley.[11] Even for species within the same genus, e.g. *Atriplex hastata* and *Atriplex rosea*, it has been shown that the latter has a lower ^{13}C content and a lower ability to produce carbon dioxide in the light.

One hypothesis[37] suggested that carbon dioxide was released by photo-respiration in both types of plant but that the additional phosphoenolpyruvate carboxylase activity present in 'Hatch and Slack' plants refixes much of the carbon dioxide liberated, preventing its escape to outside the leaf. Goldsworthy[54] measured the affinity for CO_2 of the photosynthetic system of tobacco, maize and sugar cane leaves. The concentration of carbon dioxide required to produce half the maximal rate of photosynthesis was essentially the same in all three species. This would argue against the view that there is a different carboxylation process controlling the rate of photosynthesis in 'Hatch and Slack' plants. Probably C^4 plants photorespire in the same way as C^3 plants: but their relative activity in releasing two carbon compounds from the Calvin cycle may be reduced even at high concentrations of oxygen.

The site of photorespiration

Tolbert et al.[132] succeeded in isolating from leaves of spinach, sunflower, tobacco, pea and wheat, microbodies called peroxisomes which probably occur closely adjacent to the chloroplasts in vivo. The yield from maize or sugar cane was much lower. Particles of different sizes can be separated from leaves ground in sucrose solutions and centrifuged through a series of layers of solution of different density superimposed upon each other in a centrifuge tube. For example, five suitable layers can be formed from quantities of 2.5M, 2.0M, 1.8M, 1.5M and 1.3M sucrose. A small quantity of precipitate sedimented between 100 and 6000 **g** from the first grinding of a leaf is placed on top of a series of such layers forming a density gradient and the whole centrifuged between 40 000 and 110 000 **g** for 3 hours. The particles which remain suspended in the layer of sucrose of approximately 1.9M contain a large part of the glycollic oxidase and catalase activity of the leaf. Relatively few chloroplasts or mitochondria are present in this layer. The particles present are approximately 0.5 to 1.0 μm in diameter, contain dense granular contents, are bounded by a single membrane[44] and are similar to peroxisomes from liver or kidney[30] or glyoxysomes from germinating seed endosperm. The particles also contain the enzyme catalysing the oxido-reduction of glycollate/glyoxylate using NADP (although the enzyme requiring NAD is absent), NAD/malate dehydrogenase and some transaminases. Phosphoglycollate phosphatase is not present in the peroxisomes but is confined to the chloroplasts suggesting a separation of some of the steps of the 2C path between chloroplasts and peroxisomes.[130] An enzyme concerned with the conversion of glycine to serine has been isolated from leaves of higher plants with good activity. Present indications are that it is present in the mitochondria and neither in the chloroplasts nor the peroxisomes. Thus according to present evidence the formation of serine from the intermediates of the photosynthetic cycle involves enzyme

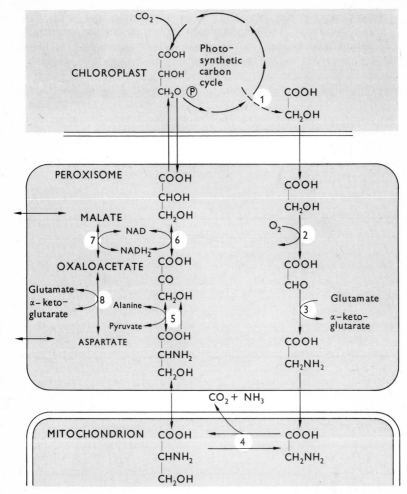

Fig. 3.6 Reactions and location of enzymes of the glycollate pathway.
1. P-glycollate phosphatase.
2. Glycollate oxidase which catalyses the oxidation of both glycollate and glyoxylate.
3. Glyoxylate-glutamate amino-transferase. A glyoxylate-serine amino-transferase.
4. A group of enzymatic reactions collectively referred to as serine synthetase or glycine decarboxylase.
5. Hydroxypyruvate-alanine transaminase.
6. Hydroxypyruvate reductase.
7. NAD malate dehydrogenase.
8. Aspartate-α-ketoglutarate amino-transferase.
(After Tolbert, N. E. (1971). *A. Rev. Pl. Physiol.*, **22**, 45.)

systems present in the chloroplast in the peroxisomes and in the mitochondria (see Fig. 3.6). The carbon dioxide evolved in photorespiration is
liberated at least in part from the mitochondria after passage of a two
carbon compound out of the chloroplast and through the peroxisome.
Recent work indicates that the conversion of glycine to serine is accompanied by a phosphorylation of ADP to form ATP in the mitochondrion.

4

The Contribution of the Chloroplast

Carbon dioxide fixation by isolated chloroplasts

After the pioneer studies of R. Hill,[59] chloroplasts could be isolated from photosynthetic tissue and shown to retain the ability to catalyse certain biochemical processes. Arnon and his collaborators[1] from 1950 onwards showed that chloroplasts isolated from spinach leaves as broken 'grana' would continue to fix carbon dioxide, but the rate was barely measurable except in the presence of a 'chloroplast extract'. Chloroplasts were extracted in the usual way, in an aqueous medium, and the liquid, from which they had been centrifuged, concentrated and added back to the preparation. However, it was also shown that the 'grana' could be pre-illuminated and if the extract was added subsequently in the dark, in the presence of $^{14}CO_2$, fixation was still observed. Thus the reaction between 'grana' and light could be separated in time from the subsequent reactions which could effect CO_2 fixation in the dark; the illumination could be omitted if ATP and reduced pyridine nucleotide were added to the grana preparation. In Chapter 2 it was seen that the essential step of the photosynthetic cycle for which light is required is the reduction of phosphoglyceric acid to triose phosphate. Now with isolated grana it had been shown that addition of the two substances necessary to drive this reaction forward could replace the requirement for light, and hence that light itself had no part in the fixation of carbon dioxide. The activity of these earliest preparations for carbon dioxide fixation was extremely low. The leaf from which the chloroplasts were obtained photosynthesized at approximately 150 μmoles CO_2/mg chlorophyll/h and the highest rate observed for the isolated chloroplasts preparation was less than 1. Consequently, because the rate was so low, it was difficult to determine whether the products of fixation were the same in the isolated chloroplasts and in the intact leaf. Nevertheless fixation into phosphoglyceric acid and into triose phosphate was demonstrated.

Gibbs and his collaborators[8] confirmed the work of Arnon *et al.* and during the next few years slowly increased the rate of carbon fixation by their preparations up to 5.0. This permitted measurements of the time course of fixation and showed the existence of an initial lag period. The lag could be eliminated by addition of fructose-1,6-diphosphate, although this addition did not alter the steady rate of fixation finally attained; it served only to prime the system and establish a steady rate more quickly. Subsequently Walker[139] attempted to isolate chloroplasts as intact as possible. For this it was necessary to use plants from which the chloroplasts could be centrifuged relatively quickly. Starch grains inside the chloroplast, which virtually act as small bullets in a centrifuge and tend to cause a puncturing of the membrane of the chloroplast, needed to be avoided. By using pea plants grown under low light intensities and at low temperatures, preparations were made which were relatively starch-free and showed minimal damage. Examination with the light microscope and phase contrast showed the preparation to consist of two components; 'whole' chloroplasts which showed a 'halo' around them, and chloroplasts in which the internal granular structure was visible. It was argued that reflection from the outer membrane of the chloroplast produced the halo whereas chloroplasts which have lost their membranes did not reflect from the surface. By making a number of preparations and comparing the proportion of chloroplasts of each type with the activity of the preparation, it was shown that those preparations with most intact membranes fix $^{14}CO_2$ at the greatest rates. Even preparations damaged as little as possible showed a marked lag in the time course of ^{14}C fixation (Fig. 4.1). If this were due to an initial deficiency of intermediates of the photosynthetic cycle addition of any component of the cycle should serve to prime the reaction equally well. In fact these compounds separated into three classes. In class 1, phosphoglycerate, phosphoglyceraldehyde and dihydroxyacetone phosphate reduced the period of the initial lag most effectively. Fructose-1,6-diphosphate was in a second class together with ribose-5- or fructose-6-phosphates, priming the reaction more slowly. Members of the third class, e.g. ribulose-1,5-diphosphate, had little effect on the lag. This difference in effectiveness may be related to the relative ease of penetration of these substances through the chloroplast envelope rather than to the true efficiency with which they 'prime' the reaction. Obviously substances which enter the chloroplast only slowly cannot prime the reaction as quickly as those which enter rapidly. Diphosphates, like ribulose, are probably relatively impermeable to the chloroplast membrane, whereas members of the first group which are smaller molecules might be expected to permeate most quickly. Use of an effective priming agent permits the measurement of higher rates

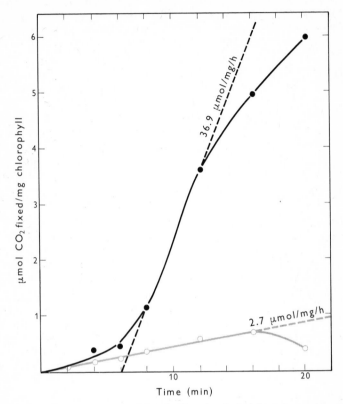

Fig. 4.1 Progress curves of CO_2 fixation by 'whole' and membrane-free chloroplasts. Reaction mixtures contained in a final volume of 0.3 cm^3 Pi, 0.25 μmol; $MgCl_2$, 0.25 μmol; $MnCl_2$, 0.25 μmol; EDTA, 0.25 μmol; Na isoascorbate, 0.50 μmol; tricine-NaOH, pH 7.5, 7.5 μmol; GSH, 1 μmol; ribose-5-phosphate, 2 μmol. Whole (●) or membrane-free chloroplasts (○) in 0.1 cm^3 suspension containing 0.028 mg chlorophyll. Reaction mixtures containing membrane-free chloroplasts also contained additional sucrose to compensate for that omitted from the resuspending medium. (From Walker.[139])

of fixation, because the chloroplasts inactivate spontaneously and the priming shortens the period of measurement. Under these conditions rates of more than 100 units were observed.

Similar conclusions were reached by Jensen and Bassham[72] who applied similar techniques to chloroplasts isolated from spinach plants which had been grown outside in the Californian sunshine. Occasionally these workers were able to observe rates in excess of the rate of the intact leaf from which the chloroplasts had been prepared. However, they did not observe any increase in rate upon addition of Calvin cycle intermediates.

The isolated grana preparations possess neither the oxidative catalysts necessary for dark respiration, nor those required for photorespiration (see Chapter 3). Since these processes confuse the study in intact tissues of photosynthesis *per se* isolated chloroplasts represent the simplest known photosynthetic system. Features affecting the kinetics of the process, e.g. the effect of light intensity or the effect of carbon dioxide concentration on the rate, can therefore be studied most simply with these preparations. However little essential difference was observed, e.g. the concentration of CO_2 required to 'half-saturate' photosynthesis (between 1 and 3 10^{-5}M free CO_2) was only slightly greater than that required to 'saturate' photosynthesis in algal cells ($5 \ 10^{-6}$M), but considerably greater than that of RuDP carboxylase *in vitro* (Chapter 2).

Jensen and Bassham[72] compared the products of photosynthesis in

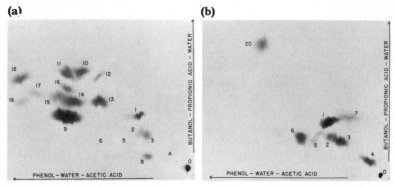

(a)   **(b)**

Fig. 4.2a Autoradiograph of $^{14}CO_2$ photosynthetic products in whole spinach leaves. Detached spinach leaves were allowed to fix $^{14}CO_2$ for 10 minutes. After killing, the leaf material was analysed by two-dimensional paper chromatography and autoradiography.

Fig. 4.2b Autoradiograph of $^{14}CO_2$ photosynthetic products in isolated spinach chloroplasts. Isolated chloroplasts were allowed to fix $^{14}CO_2$ for 6 minutes. After killing, the leaf material was analysed by two-dimensional paper chromatography and autoradiography.

The numbers in Figs 4.2a and 4.2b indicate the following compounds: 0, origin; 1, 3-phosphoglyceric acid; 2, fructose-6-phosphate; 3, sedoheptulose-7-phosphate and glucose-6-phosphate; 4, ribulose-1,5-diphosphate, fructose-1,6-diphosphate, and sedoheptulose-1,7-diphosphate; 5, ribose-5-phosphate and ribulose-5-phosphate; 6, dihydroxyacetone phosphate; 7, 3-phosphoglyceraldehyde; 8, uridine diphosphoglucose; 9, sucrose; 10, malic acid; 11, glyceric acid (?); 12, citric acid; 13, aspartic acid; 14, serine; 15, glycine; 16, glutamic acid; 17, threonine; 18, alanine; 19, glutamine; 20, glycollic acid. (From Jensen and Bassham.[72])

spinach leaves fed $^{14}CO_2$ and of the chloroplasts isolated from the same leaves. Sucrose, a major product of photosynthesis in the leaf, serine and glycine were absent from fixation in the chloroplast preparation (see Fig. 4.2). Chloroplast preparations produce largely dihydroxyacetone phosphate, the monophosphates of the hexoses and a little ribulose-1,5-diphosphate. Some glycollic acid may be formed, and when the O_2 concentration is increased glycollate accumulates to a greater extent. One report by Everson, Cockburn and Gibbs[41] that chloroplasts were able to produce sucrose has not been confirmed by other workers. Many attempts have been made to isolate the two types of chloroplasts present in maize or sugar cane leaves (Chapter 2); the rates obtained are still low and the nature of their carbon metabolism not established. Walker and Hill[140] demonstrated O_2 evolution, which was dependent on the presence of CO_2, with both pea and spinach chloroplast preparations.

The movement of carbon compounds between chloroplast and cytoplasm

As early as 1958 Tolbert[131] had determined which products of photosynthesis were retained within chloroplasts and which appeared in the chloroplast supernatant. Later both Gibbs and Walker showed that if the chloroplast reaction were allowed to take place and then the supernatant separated and added back to a new preparation of chloroplasts, the lag in fixation was decreased. Thus certain products escaped into the external medium which were capable of priming the reaction. Tolbert found that the only compound to appear outside the chloroplasts of tobacco in appreciable amount was glycollic acid. This suggested that glycollic acid readily penetrates the chloroplast membrane and may be the most permeable of all the products of photosynthesis (see also Chapter 3). If chloroplast preparations consistently fail to produce sucrose as a photosynthetic product, the possibility arises that in the leaf sucrose manufacture is partly dependent upon the enzymes in the chloroplast and partly upon enzymes in the cytoplasm. Tolbert's work suggests the possibility that glycollate may be an important component in the transfer of carbon from chloroplast to cytoplasm. Leaves photosynthesizing under normal field conditions produce a significant proportion of two carbon compounds as products and these result from the transport of carbon out of the chloroplast into the cytoplasm.

There is other evidence consistent with transport of carbon compounds between the chloroplast and cytoplasm in the intact leaf. After an intact leaf has been allowed to photosynthesize the chloroplasts can be separated from the rest of the cell. Stocking[125] first proposed a technique of isolation using non-aqueous media to prepare chloroplasts from tobacco leaves,

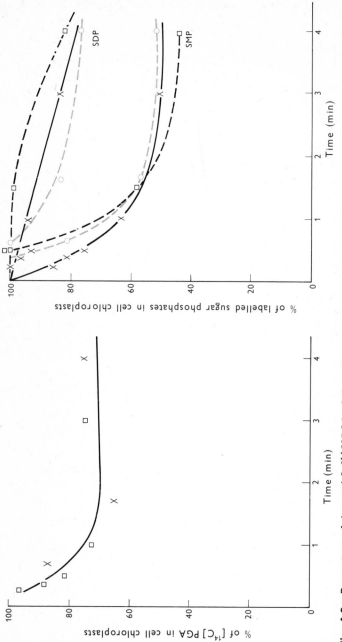

Fig. 4.3a Percentage of the total 3-[^{14}C]PGA of the cell present in the chloroplasts during photosynthesis of leaves in an atmosphere containing $^{14}CO_2$. Expt. 1 (spinach), □; Expt. 2 (spinach), x.

Fig. 4.3b Percentage of the labelled sugar phosphates of the cell present in the chloroplasts during photosynthesis of leaves in an atmosphere containing $^{14}CO_2$. SMP; sugar monophosphates, SDP; sugar diphosphates, SDP. Sugar diphosphates, SMP. Expt. 1 (spinach). ○———○; Expt. 2 (spinach), x———x; Expt. 3 (broad bean), □———□. (After Heber, U. and Willenbrink, J. (1964). *Biochem. Biophys. Acta*, **82**, 313.)

thereby hoping to minimize the loss of water-soluble compounds from the chloroplasts during isolation. Using that technique Heber et al.[58] analysed the compounds still present in the chloroplasts after isolation from the leaf (unfortunately, he found that in some plants there were substances in the cytoplasm which interfered with the analysis of the residue after extraction of the chloroplasts). Heber succeeded in showing that, when the light was turned on, the amount of ATP showed a significant rise in the chloroplast, and that when the light was turned off it fell rapidly. AMP and ADP fell in the light and showed a corresponding rise in the dark. Keys[79] repeated similar experiments with tobacco leaves using a chemical method of analysis after separation on a suitable fractionation column. The results suggest that ATP cannot readily permeate the chloroplast membrane. Both workers made a similar study of changes in NADP and NADPH, and again found no evidence that these compounds could readily permeate the chloroplast membrane.

If a leaf is fed with $^{14}CO_2$, killed at successive stages and analysed into fractions, the distribution of carbon compounds within the chloroplast and outside can be investigated. Heber found in spinach, after only a minute and a half of illumination, that PGA was no longer wholly within the chloroplasts, but approached a steady balance with approximately 65% staying within the chloroplast and 35% outside. The sugar diphosphates were relatively less permeable, the sugar monophosphates more permeable than PGA (see Fig. 4.3). These results can be compared with those from studies on isolated chloroplasts discussed earlier in this chapter which showed that PGA is relatively permeable, the diphosphates not at all, and the monophosphates somewhere in between. With tobacco leaves, Roberts, Keys and Whittingham[114] found that the first substances to appear outside the chloroplasts during illumination were glycine and serine.

Pedersen, Kirk and Bassham[111] labelled *Chlorella*, with both $^{14}CO_2$ and $H_3^{32}PO_4$, and observed the change with time in both carbon and phosphate in the PGA molecule during changes from light to dark and back to light again (Fig. 4.4). When put in the dark, the radioactivity both of the carbon and phosphorus of PGA at first showed a marked rise due to the carboxylation reaction continuing for a short time. After a longer period in the dark the radioactivity in the carbon fell but that in the phosphate was maintained. This indicates the continued formation of PGA in the dark by the process of oxidative phosphorylation associated with mitochondria. When the light was turned on a marked drop occurred in the labelling of phosphate but not of carbon of PGA. It is possible that only that part of the PGA which is in the chloroplast is rapidly converted to triose phosphate, suggesting that there must be at least two pools of PGA, that in the chloroplast which is rapidly reduced to triose phosphate, and the other in the mitochondrion

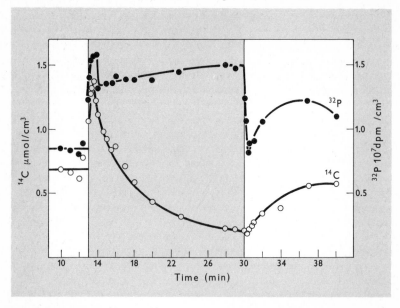

Fig. 4.4 Levels of ^{14}C and ^{32}P labelling in PGA during photosynthesis and in the dark in *Chlorella*. Ordinate: content of ^{14}C (left) and ^{32}P (right) per cm³ algae. Abscissa : minutes after addition of $^{14}CO_2$. (After Pedersen, Kirk and Bassham.[111])

which is reduced less rapidly. It is also suggested that movement can take place between the pools, if slowly, in the dark.

Consistent with all these observations, it has been suggested that certain carbon compounds may move to and fro across the chloroplast membrane forming a shuttle. This can result in an indirect transfer of either ATP or of reduced co-enzymes across the membrane whereas present evidence is that these compounds cannot themselves permeate the membrane. For example, triose phosphate might pass through the chloroplast membrane from inside to outside;[126] outside it could be oxidized to PGA with the formation of ATP from ADP and P. If the PGA could now return through the chloroplast membrane it could be again reduced with the consumption of ATP. The sequence of reactions has thus resulted in the formation of ATP outside the chloroplast and its consumption inside (see Fig. 4.5). Accompanying these reactions there will also be a transfer of reduced co-enzyme. In fact two enzymes that can catalyse this oxido-reduction have been demonstrated in plant cells. The phosphoglyceraldehyde dehydrogenase present in the chloroplast is specific to NADP and that outside the chloroplast is specific to NAD. Thus, in addition to transferring reducing power across the membrane, there is a change in the form of co-enzyme (see Fig. 4.5). A similar oxido-reduction could operate through a transfer of

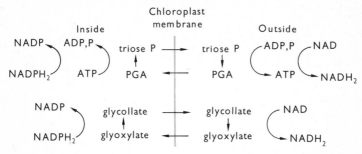

Fig. 4.5 The transfer of phosphorylated nucleotides and/or co-enzyme across the chloroplast membrane through the movement of carbon compounds.

glycollate out of the chloroplast with a corresponding return of glyoxylate; in this case the oxido-reduction would not be associated with phosphorylation.

5

Excitation and Fluorescence*

Atomic absorption spectra

An atom has a nucleus and electrons surrounding the nucleus, and when all of its electrons are in the orbitals with least energy the atom is referred to as being in the electronic ground state. If one of these electrons absorbs energy it may be ejected out into a different orbital around the nucleus with a greater energy content. Then there will be a discrete energy difference between the atom in this state and in the ground state. Only certain orbitals are possible; they cannot occur at random. Therefore for the atom to accept energy this must be offered in a unit of such a size as to allow the transition to a permitted state or it cannot be accepted.

One way for the atom to gain energy is from radiation. Radiation exists in quanta, i.e. discrete packets of energy (E) the size of which is related to the wavelength of the radiation by the equation

$$E = h\nu$$

h being Planck's constant, ν being the frequency per second. The visible region of the spectrum is due to electromagnetic waves with frequencies between 0.75 and 0.4×10^{15} cycles per second, i.e. wavelengths from 400 to 700 nm (see Appendix).

Since $$E = \frac{hc}{\lambda}$$

* Further discussion of some of the physical principles referred to in this chapter can be found in *Light and Living Matter*, Vol. 1, by R. K. Clayton, McGraw-Hill Publishing Co., New York, 1970. The Glossary (p.113) defines some terms used in this chapter.

where $c = 2.99 \times 10^{10}$ cm/s, the velocity of light in a vacuum,
     $h = 6.62 \times 10^{-27}$ ergs,

and λ is the corresponding wavelength in centimetres, a wavelength of 660 nm, or 660×10^{-7} cm, corresponds to an energy content of

$$E = \frac{(6.62 \times 10^{-27}) \times (3.0 \times 10^{10})}{660 \times 10^{-7}} \text{ ergs/quantum} = 3.0 \times 10^{-12}.$$

Since the number of molecules in a mole is given by Avogadro's number, 6.0×10^{23}, it is necessary to multiply the energy per molecule by this quantity to give the energy associated with a mole, i.e. the mole quantum or Einstein as 18×10^{11} ergs, or 185 kJ/mole (or $\simeq 2$ electron volts potential charge). Values for other wavelengths are given in Table 5.1.

Table 5.1   Energy associated with quanta of light of visible wavelengths

Wavelength nm (10^{-9} metres)	Frequency s^{-1}	Ergs/quantum	kcal/mole quantum, or Ein- stein	kJ/Ein- stein	Electron volts
400	7.5×10^{14}	4.6×10^{-12}	71.5	299	3.15
675	4.4×10^{14}	2.9×10^{-12}	42.3	177	1.84
800	3.75×10^{14}	2.5×10^{-12}	35.7	149	1.57

1 Einstein $= 6.0 \times 10^{23}$ quanta
1 kcal $= 4.1840 \times 10^{10}$ ergs $= 4.184 \times 10^3$ J
1.6×10^{-12} erg/particle $\simeq 1$ ev/particle $\simeq 23.0$ kcal/mol $\simeq 133.7$ kJ/mol
1 kcal/mol quantum $= 28.64 \times 10^3/\lambda$ nm

For a given transition in a given atom, there can be only one wavelength of radiation absorbed in accordance with the formula. If the quantum is too small or too large it cannot effect the transition. Hence for a single atom there is a simple relationship between the difference in energy content of the excited and the ground state and the wavelength of radiation which can effect the transfer. Transitions of relatively small energy content occur with infra-red radiation, intermediate with visible and large with ultra-violet absorption. The energy necessary to transfer from one atomic orbital to another is of the order of several electron volts and requires the absorption of ultra-violet light. It follows that when an atom is illuminated with a complete spectrum it will selectively absorb only those wavelengths which correspond to permitted transitions and the absorption spectrum is a series of lines given by:

$$hv = E_f - E_b$$

where E_f = energy of excited state, and
E_b = energy of ground state.

Molecular absorption spectra

Molecules are more complex than atoms since the nuclei can rotate and vibrate about each other; therefore superimposed on each molecular configuration is a range of positions of the constituent atomic nuclei, with a distribution of energy contents. However, the variation in energy content due to different vibrational (perhaps 5 to 10 kJ/mol) and even less to different rotational levels (perhaps 0.4 kJ/mol) is small compared with the difference between an excited and the ground state. An individual electron is related to more than one nucleus and a variety of orbitals about two or more nuclei exists. This produces a proliferation of energy levels and the energy of transitions between any two is decreased. Thus molecules absorb in the visible and infra-red, whereas atoms absorb only in the ultra-violet. The absorption of a quantum is so rapid an act that the nuclei of the molecule do not move and the transition is from the ground state to a corresponding energy level of the excited state. But the excited state will then also develop a range of vibrational and rotational levels. It follows that related to a single transition there are a group of wavelengths which can be absorbed corresponding to the difference between the ground state and the different vibrational states of the excited state, rather than a single wavelength. Molecules therefore exhibit an absorption band instead of an absorption line. In molecular aggregates interactions between the different molecules will cause further broadening of the absorption bands.

Chlorophyll a has two absorption bands both of which fall in the visible spectrum, with maxima at 660 nm and at 430 nm. In correspondence with these maxima the average difference in energy content between the ground state and the excited states are 180 kJ/mol and 250 kJ/mol respectively (see Fig. 5.1). The absorption spectrum for chlorophyll b, a second form of chlorophyll present in higher plants and green algae, has its two absorption maxima moved closer together, i.e. the energy content of the higher excited state is lower and that of the lower higher. Other forms of chlorophyll, c and d, occur respectively in the brown and red algae.

In addition to chlorophyll a and b, higher plants contain two accessory carotenoid pigments, β-carotene and xanthophyll. Their absorption maxima fall between those of chlorophyll, carotene having a maximum at 450–470 nm and xanthophyll at 480–540 nm. These additional pigments allow

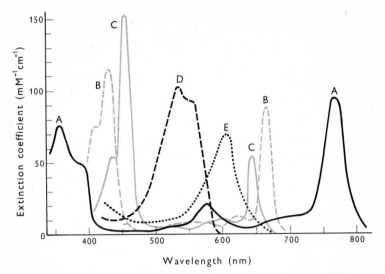

Fig. 5.1 Absorption spectra of (A) bacteriochlorophyll, (B) chlorophyll *a*, and (C) chlorophyll *b*; all in ether. (D) B-phycoerythrin, (E) C-phycocyanin; both in aqueous solution.

the plant to absorb quanta of a greater range of energy contents than it could with a single pigment molecule. In the red and blue-green algae, pigments of the class of phycobilins are also present. Phycocyanin (absorbing at 610–630 nm) and phycoerythrin (540–560 nm) are both present, but the proportion varies in different species. These two additional pigments effectively fill the region of the visible spectrum between the two chlorophyll bands. Thus the complete range of photosynthetic pigments absorb radiation through from the blue to the red end of the visible spectrum.

Fluorescence and phosphorescence

One important consideration is the life-time of the excited state. If the electron falls back very quickly, the energy absorbed by the electron is dissipated as heat; it serves only to warm the molecule. For a chemical reaction to take place, the life-time must be long enough for another molecule to collide with the excited molecule and utilize the energy of excitation. The molecule may also return from the excited to the ground state with the emission of radiation, called fluorescence. The average life-time of the molecule in the excited state will determine how many of the molecules will return to the ground state by the emission of fluorescence, and hence determine the fluorescence yield. In the excited state some of the vibrational energy may be converted to heat energy and the excited molecules will tend to sink into the lowest vibrational levels of this state; therefore the emitted fluorescent light inevitably has a smaller energy content and hence is of a longer wavelength than the exciting light ('Stokes' shift').

For chlorophyll a there is no fluorescence emission corresponding to the highest excited state. Therefore this excited state must have a very short life ($\simeq 10^{-12}$ s); it dissipates energy as heat changing from a lower vibrational level of the highest excited state to a higher vibrational level of a lower excited state. In this way absorption of blue light rapidly results effectively in the same excited state as if absorption had been in the red. The lower excited state emits fluorescence with a characteristic waveband maximum at 690–700 nm (see Fig. 5.2). Since the life of this particular

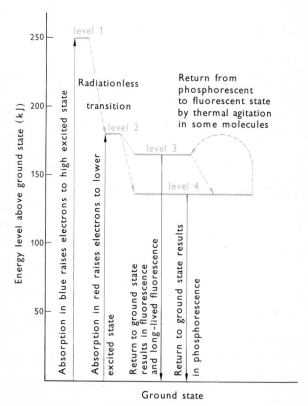

Fig. 5.2 Energy levels of the chlorophyll molecule.

excited state is sufficient to allow fluorescence ($\simeq 10^{-8}$ s) it may also allow collision with other chemical molecules, thereby initiating chemical reactions. From the wavelength of the fluorescent light the average energy of the excited state can be calculated to be 168 kJ/mol. Chlorophyll *b*, which also has two absorption maxima, emits fluorescence at a single waveband slightly shorter than that of chlorophyll *a* (maximum 670 nm). The fluorescence yield is about the same, and hence the excited state must have a similar life-time to that of chlorophyll *a*. Carotenes do not emit fluorescence and neither do xanthophylls. The phycobilins have a higher fluorescence yield than the chlorophylls and might also be expected to initiate photochemical change.

The other group of red and blue pigments in plants, the anthocyanins, which are largely responsible for autumn colours of leaves and the colour of many flowers, do not occur in the chloroplast and have no connection with the process of photosynthesis.

In addition to fluorescence, certain excited molecules continue to emit light as phosphorescence after the exciting light has been turned off. Such molecules must possess an excited state which has a longer life-time than the fluorescent state, perhaps of the order of 10^{-3} s. Such a state is often a triplet excited state in which the spin of an electron has become reversed during excitation; the shorter-lived excitation states referred to previously are singlet excited states in which the spin of an excited electron remains antiparallel to that of its ground state partner. Transitions from the ground state to excited triplet states are much less probable than to excited singlet states and entry of a molecule into a triplet state is more probable from an excited singlet state than from the ground state by direct excitation. From such a long-lived excited state some molecules continue to return to the ground state emitting radiation after the exciting light is turned off. Phosphorescence of chlorophyll *b* corresponds to an energy state of 146 kJ/mol. The difference between the fluorescent and the phosphorescent states is thus only 22 kJ/mol. Hence certain molecules can transfer spontaneously by thermal agitation from the phosphorescent to the fluorescent state thus populating this state after excitation has ceased. When this happens the radiation emitted corresponds to fluorescence rather than phosphorescence and is then designated long-lived or 'slow' fluorescence. This last phenomenon has been demonstrated only for chlorophyll *b* under very special conditions; it has not yet been demonstrated for chlorophyll *a* or for any other plant pigment.

The photosynthetic pigments *in vivo*

The properties of the pigments discussed so far refer to their behaviour when they are extracted from the plant in an organic solvent, e.g. ether or

acetone. It may be assumed that these properties are not greatly changed for
the pigments in the plant (Fig. 5.3). If the individual absorption spectra of

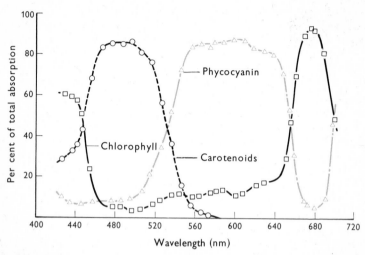

Fig. 5.3   Relation to wavelength of the percentages of total light absorbed by
Chroococcus sp. due to chlorophyll, carotenoids and phycocyanin respectively;
estimates based on measurements with extracted pigments. (After Emerson, R.
and Lewis, C. M. (1942). *J. gen. Physiol.*, **25**, 579.)

the pigments extracted from a plant are added in appropriate proportions,
their sum approximates to the absorption spectrum of the plant itself. In
the plant the absorption bands are broadened (half bandwidth 30 nm compared to 18 nm in solution) and the absorption maxima moved to longer
wavelengths, i.e. 680 as compared with 660 nm for chlorophyll *a*. It is unlikely that the energy states are sufficiently modified to make observations
on pigments in solution irrelevant. Thus the contribution of each pigment to
absorption at a given wavelength may be calculated (Fig. 5.3). The fluorescence spectrum of chlorophyll *a in vivo* has a maximum at 685 nm with a
smaller maximum at 740 nm. But the yield of chlorophyll *a* fluorescence
in living cells is 3–5%, whilst that *in vitro* is of the order of 30%. Accord-
ingly the life-time *in vitro* is of the order of 5–15 ns whereas *in vivo* the
life-time is more nearly of the order of 1–1½ ns. [142,88] One explanation of
this difference could be to suppose that a significant portion of chlorophyll
a in vivo, perhaps as much as two-thirds, does not contribute to fluorescence.[15] Furthermore, fluorescence from chlorophyll *a* markedly decreases as the excitation wavelength is altered to the long wavelength side
of the red absorption band. This 'red drop' can be interpreted in terms of

the existence of a special form of chlorophyll absorbing at longer wavelengths which is only weakly fluorescent.[27] Again *in vivo* when fluorescence
is excited by polarized light a weak polarization of the fluorescence is
observed; the emission of polarized fluorescence shows a maximum at
about 718 nm whilst that of the depolarized fluorescence is 685 nm. This
suggests that part of the chlorophyll *a* molecules have a different orientation from the bulk of the molecules.[91] The decay curve of chlorophyll *a*
fluorescence *in vivo* has been studied by a number of authors. When the
logarithm of the fluorescence intensity is plotted against time two straight
lines are obtained: one corresponding to a half life of 1–2 ns, the other to a
half life of 4–5 ns.[106] Hence there are many indications that chlorophyll *a*
may exist *in vivo* in two different forms probably relating to a difference
either in chemical bonding or physical state.

All attempts to extract different forms of chlorophyll with organic solvents from plants have so far failed; all procedures produce only a single
pigment molecule with a maximum absorption at 660 nm in organic solution. Thus the different forms of chlorophyll postulated must exist only *in
vivo*. For example, that part of the chlorophyll absorbed onto lipids in the
chloroplast might be functionally different from that part associated with
the hydrophilic proteins. Some more direct evidence for the existence of different forms of chlorophyll *in vivo* has come from detailed studies of
absorption spectra.[18] When the absorbance (D) is plotted against wavelength (λ) for a single absorption band, a curve symmetrical about the maximum is observed. If then the differential of the absorbance with respect to
wavelength $d(D)/d\lambda$ is calculated it is clear that at the wavelength for
maximum absorption $d(D)/d\lambda$ must have a minimum. With a mixture of
two substances the relationship between

$$\frac{d(D)}{d\lambda} \quad \text{and} \quad \lambda$$

would be more complex; it would be the sum of the two separate derivative
absorption spectra and would not necessarily show a minimum at the
wavelength of maximum absorption of either component. The derivative
curve observed for a plant *in vivo* will be given by summing all the curves
for each individual component. Experimentally French and Brown found
that a minimum of three component curves was required to fit the derivative absorption curve for an alga representing components with absorption
maxima at 670, 682 and 692 nm, referred to as chlorophyll a_{670}, chlorophyll
a_{680} and chlorophyll a_{690} (see Fig. 5.4). Similar results were obtained with
a number of species of green algae and when the cultural conditions for the
algae were varied the proportion of the different components also varied.
For a young culture of *Euglena* there was an excess of the component

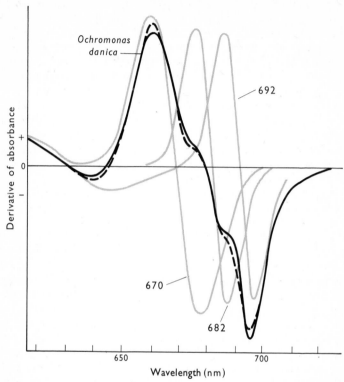

Fig. 5.4 The derivative of absorbance for whole cells of *Ochromonas danica*. This curve can be analysed in terms of three components which have symmetrical absorption curves. The addition of the derivative absorption spectra for the three components (– – – –) is seen to correspond closely to the measured spectrum. (After Brown and French.[18])

absorbing near 670 nm but as the culture 'staled' the proportion of the component absorbing at 695 nm increased.

The leaf of a higher plant grown in the dark is pale green but when put into the light rapid chlorophyll synthesis takes place. The pigment present in the etiolated leaf, protochlorophyll, has an absorption maximum at 645 nm. After the leaf is given only a short exposure to light, a pronounced chlorophyll peak at 684 nm is observed.[122] If following the initial light period the leaf is placed again in the dark, the spectrum shifts rapidly to a maximum at 673 nm and after some hours two maxima at 673 and 683 nm are finally obtained. This again suggests the existence *in vivo* of different chlorophyll forms.

When the fluorescence of chlorophyll *a* in solution is measured the fluorescence intensity is proportional to the exciting intensity over a wide range of intensities, i.e. the quantum yield of fluorescence is constant. *In vivo* however, the fluorescence yield of green algae at higher intensities of illumination (i.e. intensities which saturate photosynthesis) is twice that at lower intensities (where the rate of photosynthesis is proportional to intensity). The fluorescence yield is also increased when the concentration of carbon dioxide is decreased or certain inhibitors of photosynthesis added.[89] These effects can be interpreted in terms of a competition between the emission of fluorescence and the use of excitation energy for photosynthesis. But the effects observed are more complicated than this. When photosynthetic cells are first illuminated following a dark period the fluorescence intensity rises rapidly to an initial level but then subsequently rises to a new level reached within 1–2 s, after which it subsequently falls to a steady state value. The exact shape of the curve as a function of time depends on the experimental conditions and the organism studied. Lavorel[90] suggested that the curve should be analysed in terms of two components, one invariant with time with a maximum emission near 717 nm, and a variable component with a maximum near 685 nm. The variation in the second component giving fluorescence is attributed to a cellular constituent which is oxidized in the dark but becomes reduced as a consequence of a photochemical process. In the oxidized form the substance is presumed to act as a quencher of fluorescence and has therefore been designated Q. Its chemical identity is not known; it may be a form of plastoquinone or a particular molecule in association with part of the plastoquinone of the cell. This type of mechanism has been elaborated independently by a number of workers including Duysens, Kok and Joliot.[36,73] By comparison photosynthetic purple bacteria (which do not evolve O_2 in photosynthesis as discussed in Chapter 6) show a simpler time course of fluorescence and show no 'red drop'; there is no evidence in the bacteria of two bacteriochlorophyll components, a fluorescent and a non-fluorescent.

In a later chapter it is shown that two photochemical systems are probably involved in photosynthesis in green plants and it is now believed that one of them (system II) causes reduction of Q, the other (system I) oxidation. If a leaf has been previously illuminated by red or far red light, the yield of fluorescence due to a following excitation by shorter wavelengths is greatly reduced. Again it was observed that the fluorescence at 715 nm was virtually independent of the pre-illumination of the sample, but that the fluorescence at 685 and 693 nm showed marked fluctuations.[21] The effect of a poison of photosynthesis such as DCMU, 3-(3,4-dichlorophenyl)-1,1 dimethylurea, is due largely to an increase in the 685 nm component only.[86] It may be concluded that fluorescence *in vivo* originates from two forms of chlorophyll, one emitting at a maximum near 685 nm (with a minor band at 640 nm) and the other near 720 nm. The 685 nm band probably originates from the form to be discussed later as system II pigment, whilst the 720 nm band originates mainly from system I; a minor part of the 685 nm band could however arise from system I also.

Fluorescence and energy transfer

Another property of the pigments in the plant, which is also exhibited when they are mixed in high concentration in organic solvents, is that the fluorescence of the mixture is different from the sum of the emissions of the separate components. Chlorophyll b and chlorophyll a in ethyl ether absorb at 640 and 480 nm and 660 and 440 nm respectively. Chlorophyll a will emit fluorescence about 690 nm and chlorophyll b at about 670 nm. Consequently in a dilute solution containing both chlorophyll a and chlorophyll b excitation at 640 nm should give fluorescence only at 670 nm since when no light is absorbed by chlorophyll a, no chlorophyll a fluorescence will occur. But with a concentrated mixture of the two pigments, although light which excites chlorophyll b alone is used, fluorescence characteristic of chlorophyll a excitation is emitted and it is possible to suppress completely the fluorescence due to chlorophyll b (see Fig. 5.5). Thus, energy transfer must have taken place between molecules of the two different chlorophylls in order that the fluorescence of the second species can be sensitized by excitation of the first. The energy transfer has taken place by resonance in which the de-excitation of b generates an electromagnetic field which in turn excites a. The efficiency of transfer can be very great and is not consistent with a mechanism in which b emits fluorescent quanta which are re-absorbed by a; the time of resonance transfer is very short. Such transfer can also occur in an aggregate between molecules of the same species. This can be demonstrated if the exciting light used is polarized. If no energy transfer takes place the fluorescence emitted should also be polarized assuming the life-time is sufficiently short not to permit rotation. But if energy transfer takes place through a series of randomly orientated molecules the polarization will be lost. From such observations with chlorophyll a *in vivo*, Teale[129] calculated an average transfer through 275 molecules before emission. The value will be inaccurate if the chlorophyll molecules are orientated within the chloroplast and not arranged at random; earlier in this chapter it was suggested that two chlorophyll forms emit *in vivo* with different orientations with respect to each other.

In red algae illuminated by monochromatic light at about 550 nm excitation will be largely of phycoerythrin yet a fluorescence emission characteristic of chlorophyll a is observed.[33] Again, the yield of chlorophyll a fluorescence in the green alga *Chlorella* is largely independent of the wavelength of exciting light. It follows that all pigment molecules, whether

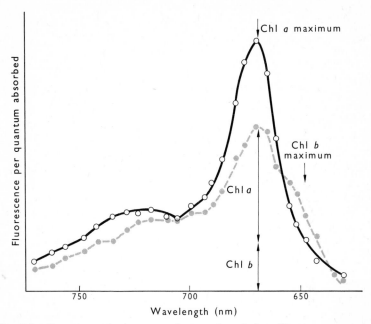

Fig. 5.5 Energy transfer between pigments in solution. Fluorescence spectra for chlorophylls in ether solution when irradiated with a wavelength of 429 nm (70 per cent absorption due to chlorophyll *a*) and with 453 nm (95 per cent absorption by chlorophyll *b*). The vertical line at 670 nm indicates that in the second case a large portion of the fluorescent light is characteristic of chlorophyll *a*, whereas chlorophyll *b* was the pigment excited: partial transfer of energy from chlorophyll *b* to chlorophyll *a* has taken place. In the living cell the pigments are more concentrated and transfer is practically complete. ○, Exciting light 429 nm; ●, exciting light 453 nm. (After Duysens.[33])

xanthophylls or phycobilins or chlorophylls other than chlorophyll *a*, transfer the energy which they absorb to chlorophyll *a* and the end effect is to excite chlorophyll *a*. Hence the role of these pigments is accessory; they are only secondary absorbers and do not emit fluorescence or probably initiate any photochemical reactions themselves. In the plant, energy transfer has been shown to be very efficient; in excess of 90% of the energy absorbed is transferred in most of the cases so far studied. One exception is that the energy absorbed by the carotenes is only poorly transferred if at all. During dissipation of excitation energy largely as vibrational energy, a state of the donor molecule is attained which resonates with a vibrational state of the acceptor molecule. The molecule to which the energy is to be transferred must have a lower energy content than the molecule from which transfer takes place, i.e. the fluorescence band of the acceptor molecule must overlap the absorption band of the donor. The greater the overlap of the fluorescence and absorption bands and the closer together the two molecules concerned the greater the efficiency of the energy transfer. The transfer may be compared to the transfer of vibration from a vibrating to a static tuning fork provided the two forks are tuned and placed close together.

Action spectra of photosynthesis

Similar conclusions can be reached by comparing the amount of photosynthesis produced by equal energy supplied at different wavelengths of exciting light, i.e. by studies of the action spectrum.* If chlorophyll *a* were the only pigment which could sensitize photosynthesis and there was no energy transfer, the action spectrum for a green plant would correspond

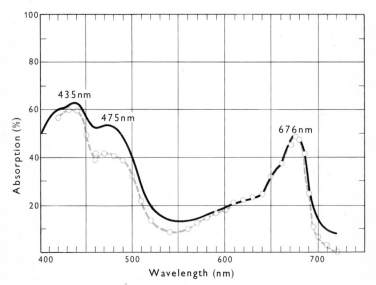

Fig. 5.6 The absorption spectrum for a cell suspension of *Chlorella pyrenoidosa* (————) compared with the photosynthetic action spectrum (----) (After Haxo, F. T. (1960). In *Comparative Biochemistry of Photoreactive Systems*, Ed. Allen, M. Belle. Academic Press.)

* An action spectrum may be defined more precisely as the reciprocal of the intensity (measured in Einsteins cm^{-2} s^{-1}) required to bring about a certain response as a function of the wavelength of the exciting light.

to the absorption spectrum of chlorophyll *a*, the only useful pigment. If chlorophyll *b* transfers its energy to *a*, then the action spectrum would extend to include *a* and *b*. Hence by a careful comparison of the absorption and action spectra, the efficiency of pigments in transferring absorbed energy to photosynthesis can be observed. In the alga *Chlorella* (Fig. 5.6), a comparison of the absorption spectrum and the action spectrum shows a discrepancy only in the region where absorption by β-carotene is maximal.[40] Thus absorption by carotene is unable largely to transfer energy into photosynthesis and it will be remembered that it also sensitizes fluorescence of chlorophyll only inefficiently. By setting up a series of equations for a number of wavelengths relating the rate of photosynthesis to the relative absorption by the individual pigments, transfer coefficients for each pigment can be calculated. The calculation assumes that the pigments are homogeneously distributed within the plastid and have the same absorption properties as *in vivo*. In all cases where transfer takes place a coefficient of the order of 90% is calculated; β-carotene with a value near zero is the exception.

It may be expected that if different forms of a pigment exist *in vivo* transfer of energy could take place between them with a high efficiency. Evidence discussed earlier in this chapter indicates that transfer did not take place efficiently between the two forms of chlorophyll *a* 670 and 682. Therefore it must be supposed that they are separated physically within the chloroplast, possibly within different sub-units within the plastid. But within any one structural group transfer can take place so that ultimately the energy absorbed by the group as a whole can find its way into certain individual molecules called 'reaction' centres or 'trapping' centres. If two species of molecule are present, that form which absorbs at the longest wavelengths will trap the energy absorbed by all other species; within a group of molecules of the same species the energy could migrate between them until finally reaching a particular molecule which was, for example, complexed with an enzyme molecule. In this way a limited number of enzyme molecules which can react as electron donors or acceptors can utilize the light energy absorbed by any one of a large number of pigment molecules. Such an aggregate has been called a photosynthetic unit. An individual chlorophyll molecule will absorb on average under weak natural light conditions one quantum only every few minutes; but if the energy of a number of pigment molecules contributes to a single reaction centre that can be energized every fraction of a second.

A special case was observed in certain red seaweeds and in blue-green algae which contain phycoerythrin and phycocyanin. In the region where chlorophyll predominantly absorbs (between 660 and 680 nm), little photosynthesis could be observed. But by contrast where the phycobilins absorb, maximal photosynthesis was observed by Haxo and Blinks[57] (see

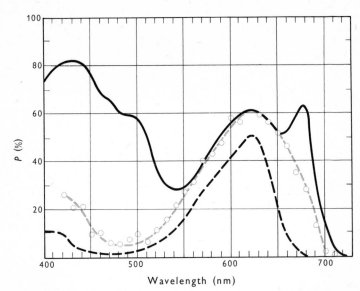

Fig. 5.7   The absorption spectrum of a cell suspension of *Cyanidium caldarium*
(———) compared with the action spectrum for photosynthesis (O----O)
The absorption due to C-phycocyanin is shown (– – – –). The action spectrum
resembles the absorption spectrum of C-phycocyanin much more closely than
that of intact cells.

Fig. 5.7). They later studied the fluorescence spectra and observed very
effective transfer from the phycobilins to chlorophyll. Again the fluorescence of chlorophyll was greater with phycobilin excitation than if the
chlorophyll had absorbed the light energy direct. This apparent paradox
will be discussed in detail in the next chapter.

6

The Physiological Evidence for Two Photochemical Reactions in Green Plants

Detailed examination of the action and fluorescence spectra of *Chlorella* for wavelengths longer than 690 nm showed an increasing difference between light absorption and photosynthetic action or fluorescence, the latter decreasing much more sharply with increase of wavelength. This phenomenon of the 'red drop' suggested the existence *in vivo* of a form of chlorophyll absorbing at the longer wavelengths which could not sensitize photosynthesis or fluorescence. In that case the accessory pigments which produce efficient photosynthesis by transfer of energy must do so preferentially to the efficient form of chlorophyll which absorbs at wavelengths shorter than 690 nm. The inefficient form of the chlorophyll which accounts for a greater proportion of absorption as the wavelength increases beyond 670 nm cannot receive energy from the accessory pigments; if it did so, such energy would be used with only low efficiency in photosynthesis. Further the two forms of chlorophyll, one capable of producing photosynthesis and one not, must be in separate compartments with little possibility of energy transfer between them.

Difference spectra

Normally an absorption spectrum is determined using monochromatic light of weak intensity (measuring light). If at the same time a second beam (the actinic light) is shone on the cuvette at right angles, the effect of illumination by the beam on the optical density of the sample can be determined (Fig. 6.1). The major problem in such measurements is an inter-

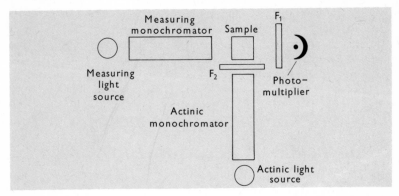

Fig. 6.1   Diagram of apparatus for determining the effect of an actinic beam on the absorbing of a sample. F_1 and F_2 are suitable complementary filters (see text).

ference in the measuring light by the actinic light. The simplest device to exclude interference is by the use of complementary filters. If for example the measurement is made in the blue part of the spectrum, and red actinic light is used, then with a suitable filter which transmits only blue light between the sample and the receiver, the direct effect of the actinic light on the measurement can be reduced to a minimum. Alternatively, the measuring beam can be made an alternating signal by the use of a rotating disc and the actinic light maintained as a steady signal. By use of a detector or amplifier which is sensitive only to alternating signals, the influence of stray actinic light on the measurement can be very considerably reduced. To avoid variations in the source with time, a double beam system is frequently used in which the measuring beam is split, one portion passing through a reference system, the other through the sample. The method can be further refined if the reference beam is passed through a comparable sample which is illuminated with actinic light of a wavelength which is known to give little photoreaction as compared with the wavelength passing through the measured sample. If two wavelengths can be chosen reasonably close together, such a device compensates for changes in scattering introduced by the sample. Errors also arise if the actinic light causes fluorescence at a wavelength near that of the measuring beam. In practice the best results can be obtained by using the double beam system and alternating the measuring signal. The changes measured can then be of the order of $\Delta D \simeq 0.005$, i.e. less than 1% of the basic signal. Even so, it is necessary to interpret the data with a full regard to the effects of artefacts which can influence appreciably such small changes.

The first measurements of this type were made by Duysens[34] who concentrated on the changes in absorption resulting from a steady state of illumination. In subsequent work, both Witt[147] and Kok[82] used an actinic flash rather than continuous illumination. With an extremely brief flash of the order of 10^{-5} s, measurements could be made subsequent to the flash of the absorption changes which it had induced. By measuring both the rise time of the absorption change and the decay time, it is possible to analyse further a situation where two different absorption changes superimpose at the same wavelength.

The absorption curve for cytochrome c shows maximum absorption about 550 nm (α band) and 430 nm (γ band); when cytochrome c is oxidized the sharp α absorption band is replaced by a weaker and more general absorption and the γ band is displaced towards shorter wavelengths and diminished in intensity. Chance[23,24] used this change in optical density to study the oxido-reduction changes of cytochrome during respiration. The same principle can be applied to photosynthetic tissues. When instead of adding a reducing agent light is put on, if in the course of photosynthesis cytochrome is reduced, the absorption at 430 nm should show a significant decrease. When the measuring beam is made to scan right through the spectrum, to observe changes in absorption at a complete range of wavelengths, a difference spectrum is observed.

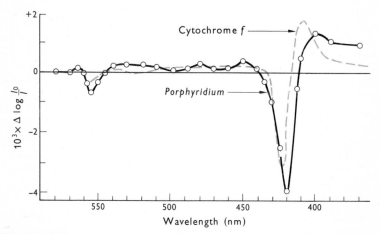

Fig. 6.2 Difference spectrum of *Porphyridium cruentum*, i.e., absorption spectrum in light, minus that in darkness. The difference spectrum of cytochrome *f*, oxidized minus reduced, was obtained from data of Davenport and Hill[28] (After Duysens *et al.*[35])

Duysens *et al.*[35] first measured the difference spectrum resulting from steady state photosynthesis in the red alga *Porphyridium*. The difference spectrum obtained for chemical reduction of cytochrome *f in vitro* corresponded quite closely to the difference spectrum resulting from illumination of a suspension of *Porphyridium* cells (Fig. 6.2). It follows that cytochrome *f* probably undergoes changes in oxido-reduction during the course of photosynthesis. Subsequent work has shown that *Porphyridium* has the simplest difference spectrum of all organisms so far studied. With *Chlorella*, instead of getting a relatively simple difference spectrum, at least five regions of significant change were observed in the visible part of the spectrum. No one has yet been able to interpret this complex difference spectrum completely in terms of known biochemical intermediates.

To analyse the difference spectrum further, Witt[104] used illumination with brief intense flashes, and attempted a kinetic analysis of the difference spectrum resulting from the cumulative effect of a number of flashes as a function of time (see Fig. 6.3). Witt classified type O changes induced by

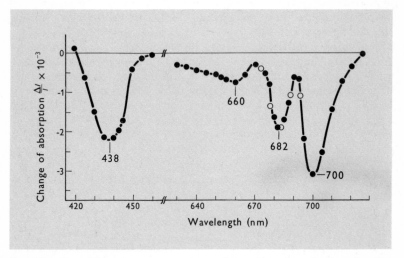

Fig. 6.3 The difference spectrum for a light chloroplast fraction (smaller particles) measured with a repetitive pulse technique at a light time of 2×10^{-2} s, as a function of wavelength. (After Döring, G., Bailey, J. L., Kreutz, W., Weikard, J. and Witt, H. T. (1968). *Naturwissenschaften*, **55**, 219.)

the shortest type of flash (3×10^{-5} s) which showed a characteristic increase at 530 nm (see Table 6.1). Witt first postulated that this absorption change was due to the formation of a triplet state of chlorophyll *in vivo*, but this is now doubtful. Only when the flash is very short and the temperature kept as

Table 6.1 Components of the difference spectra of green plants according to Witt et al.[147]

Rise time (s)	Decay time (s)	Characteristic wavelength change (nm)	Substance believed responsible
10^{-8}	10^{-6}	430, 1460, $-490, +520$	Carotene
10^{-8}	10^{-2}	$-438, -700$	Chlorophyll a_I (P_{700})
10^{-8}	10^{-4}	$-435, -682$	Chlorophyll a_{II} (P_{680})
10^{-3}	10^{-2}	-263	Plastoquinone
10^{-4}	10^{-2}	$+407, -424, -555$	Cytochrome f

low as that of liquid nitrogen, can spectral evidence for formation of a triplet state be observed; under normal conditions of photosynthesis there is no evidence for triplet state formation. A longer type 1 flash (10^{-4} s) produced a characteristic decrease at 485 nm, and an accompanying increase at 520 nm. Such changes were destroyed by heating above 60°C. Type 2 changes, induced by a flash of 10^{-2} s duration, produce a decrease at 420 nm and also at 475 and 515 nm. When chloroplasts were used after extraction with petrol ether, which was considered to extract plastoquinone, the last two peaks were lost. The activity was restored by adding back plastoquinone. Hence it was suggested that plastoquinone was involved in photosynthesis.[3] Plastoquinone itself does not show a difference spectrum in this region of the spectrum but an unidentified complex of plastoquinone in the plastid is postulated which changes its spectrum as a result of changes in the associated plastoquinone. Type 2B changes, resulting from the longest flashes of all, produce an increase in absorption at 515 nm and a concomitant decrease at 475 nm. This is the largest response in this region of the visible spectrum and increases progressively with increase in intensity of actinic light. The changes were at one time attributed to a single biochemical entity, but now it is generally believed that the change may be due to several as yet unidentified components possibly including chlorophyll *b* and carotenoids.

Kok[81] used intense flash illumination both with algae and isolated chloroplasts and showed a significant absorption change in the red part of the spectrum at 700 nm associated with a smaller change at 430 nm. This could not be related to any known substance. He therefore called the substance responsible for this change in difference spectrum P_{700}. Kok considered P_{700} to be a form of chlorophyll *a* occurring *in vivo* which is reversibly bleached by light. By addition of redox agents of different potential and observation of their effect on the difference band at 700 nm, it was shown that the bleaching of P_{700} related to its oxidation and the potential was determined as about $E'_0 + 0.400V$.* It is a one-electron redox carrier. Its probable concentration is very small, perhaps as little as one molecule for each 200 to 1000 molecules of chlorophyll a, assuming a molar extinction coefficient in the reduced form the same as that of chlorophyll. Many workers consider that P_{700} is a special part of the chlorophyll a present and may be that part which gives an absorption maximum at 692 nm. The decay time of the 700 nm difference is of the order of 10^{-2} s. If chloroplasts are extracted with an acetone 70–72%

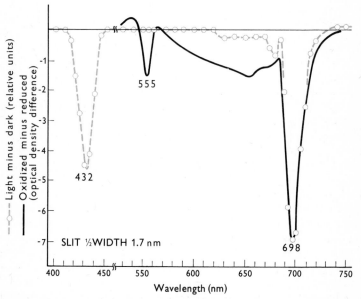

Fig. 6.4   Full line : oxidized minus reduced difference spectrum. Broken line : light minus dark spectrum. Both measured with an acetone extracted suspension of chloroplast particles which has retained much of the P_{700}. (After Kok, B. (1961). *Biochim. Biophys. Acta*, **48**, 527.)

v/v water mixture a great deal of the chlorophyll a is extracted but less of the P_{700}. After such treatment the absorption changes due to P_{700} remain but no change attributed to cytochrome f is observed (Fig. 6.4).

In a similar manner if chromatophores isolated from photosynthetic

* E'_0 is the standard oxidation-reduction potential at pH 7.0 and 25°C referred to the standard hydrogen electrode. It is a measure of the standard free energy for reduction in the reaction $A + H_2 \rightarrow AH_2$ per electron transferred.

bacteria are treated with detergent most of the bacteriochlorophyll can be removed and only a special part called P_{890} remains.[138,25] This pigment is strongly bleached by light due to its oxidation; accompanying this change there is also a shift of the absorption band at 800 nm to shorter wavelengths. By the use of laser beams[110] it has been shown that this change takes place within 10^{-6} s at 37K. Both P_{700} and P_{890} may be the substances which are the first to be oxidized by the excited pigment molecule. The term reaction centre has been used to define the particular molecular grouping at which excitation energy is first converted to chemical energy.

Enhancement spectra

Again by means of the same type of experimental arrangement whereby a reaction vessel can be simultaneously illuminated by two monochromatic beams, the effect on the production of oxygen by photosynthesis can be measured. If the wavelength of one beam is varied through the spectrum, the addition of the second beam of light of constant wavelength would be expected to increase the rate of photosynthesis, and produce an action spectrum uniformly enhanced at all wavelengths. The additional photosynthesis resulting from the presence of the second beam as a function of wavelength of the first beam is called the enhancement spectrum. Special attention paid to the region of far red absorption in green algae shows that when a second beam of blue light is added photosynthetic action extends out to wavelengths beyond 690 nm. The blue light, even though of relatively low intensity, has produced photosynthesis from far red light in a region where little or none occurs when illumination is by the far red light alone. This phenomenon discovered by R. Emerson[39] is called the Emerson enhancement effect. Photosynthesis is enhanced in this region by the addition of light of shorter wavelengths to a far greater extent than the arithmetic sum of the photosynthesis produced by the two beams of light given separately.

By varying the wavelength of the second beam it was found that when any of the accessory pigments are simultaneously excited, together with far red excitation of chlorophyll, the photosynthetic activity in the far red is greatly enhanced (see Fig. 6.5). This suggests that the two forms of chlorophyll referred to at the beginning of this chapter may both be involved in photosynthesis under special conditions. The pigment absorbing in the far red (pigment system I) is unable to work alone; the energy absorbed by it is effective only when the other system, called pigment system II (absorbing at shorter wavelengths), is also excited. Hence the longer wavelength system must react sequentially with some product of the shorter wavelength system. Unfortunately it is theoretically impossible to do the opposite, i.e. to excite the shorter wavelength system without also exciting the longer wavelength system, since absorption by chlorophyll a in all its

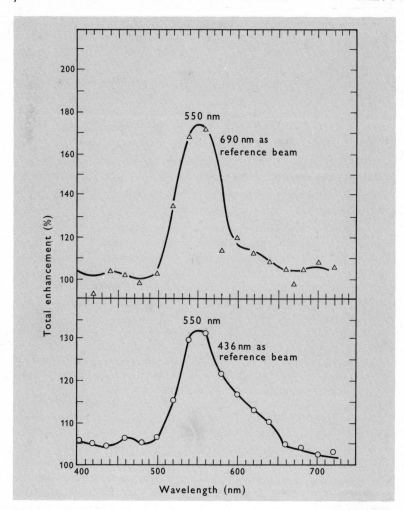

Fig. 6.5 The action spectra for photosynthesis enhancement resulting from supplementation of either far red light (wavelengths greater than 690 nm) or of blue light (436 nm) in the red alga, *Porphyra perforata*. (After Fork (1963). *Photosynthetic Mechanisms of Green Plants*. Nat. Acad. Sci., Washington, Publ. 1145, p. 352.)

forms probably extends right through the shorter wavelengths. It is only because as absorption extends into the far red the number of forms of pigments absorbing decreases one by one that the pigment absorbing furthest into the red can be excited alone.

Characterization of the two systems

The effect of the two photochemical systems can be most clearly separated in time when their effect is studied through the difference spectra they induce. Duysens et al.[35] showed that if light absorbed by pigment system I was given to red algae, cytochrome f became oxidized, e.g.

$$\text{Chl. cyt} \xrightarrow{h\nu} \text{Chl.}^{\star} \text{cyt} \longrightarrow \text{Chl.}^{+} \text{cyt}^{-}$$

where Cyt^{-} indicates cytochrome which has gained an electron and become oxidized and Chl^{+} chlorophyll which has lost an electron. The reduced chlorophyll Chl^{+} must regain its electron ultimately from some other molecule which itself becomes reduced. When subsequently light absorbed by system II was given, the return of the oxidized cytochrome to the reduced state was accelerated (see Fig. 6.6), the electron donated by

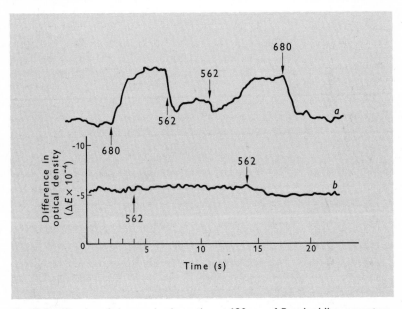

Fig. 6.6   Kinetics of changes in absorption at 420 nm of *Porphyridium cruentum*. Upward or downward pointing arrows indicate that the actinic light is switched on or off. Actinic light of 680 nm (mainly absorbed by system I) caused an oxidation of an f type cytochrome (decrease in absorption at 420 nm). When in addition a further beam of light of 562 nm (predominantly absorbed by system II) was admitted the absorption at 420 nm was restored indicating reduction of the cytochrome. The intensities of actinic beams were 44×10^{-11} and 32×10^{-11} Einsteins cm^{-2} s^{-1} at 680 and 562 nm respectively. (Courtesy of Professor L.N.M. Duysens.)

system II ultimately coming from water. Also, the Emerson enhancement effect can be observed when the two light beams are given not simultaneously but successively. Myers and French[108] showed that in *Porphyridium* oxygen evolution from a green flash was enhanced when it was preceded by a red flash. This suggested that some intermediate produced by system II had survived during the period between the two flashes. At temperatures near $0°C$ the half time of survival was of the order of several seconds. Myers and French claimed that the reverse phenomenon could not be observed, i.e. oxygen evolution from a red flash was not increased following a previous green flash. This indicated that certain products of system I have a significant half life whereas there are no long-lived intermediates resulting from photochemical system II.

From photosynthetic enhancement studies there is evidence that system I in green algae and leaves contains the special form of chlorophyll *a* absorbing at longer wavelengths, identified by absorption studies (see Chapter 5) as chlorophyll a_{690} and probably some of the chlorophyll a_{680} component also. System II pigments include certainly the largest part, if not all, of chlorophyll a_{670}, all the accessory pigments (including chlorophyll *b*), and possibly also some chlorophyll a_{680}. As the wavelength of light varies the relative amount of energy absorbed by the two systems varies; beyond 680 nm 'excess' energy is absorbed by system I relative to that absorbed by system II. The excess is wasted and the 'red drop' in photosynthesis occurs. When the reverse situation occurs, i.e. the excess is in system II, excess energy could be transferred from system II to I (although the reverse would be thermodynamically impossible). Such a 'spill-over' hypothesis was proposed by Myers[107] as an alternate to the completely independent operation of the two systems as formulated in the 'separate package' hypothesis. Experimental data are still inadequate to establish one view rather than the other. Some evidence suggests that illumination with system II light for 2–3 minutes may increase the proportion of pigment absorbing in system I and vice versa for system II; this suggests the movement of pigment molecules between the photochemical units rather than transfer of energy. By studying the action spectra for the various absorption changes of the difference spectrum it can be shown that the change at 433 nm is greatest with stimulation of system I and that at 515 nm with excess stimulation of system II (see Fig. 6.7).

In *Porphyridium* and in blue-green algae the phycobilins are the predominant pigment of system II. System I must contain most of the chlorophyll. Thus in these groups of organisms the 'red drop' of green algae extends over a wider range of wavelengths so that absorption within the whole chlorophyll *a* region is relatively inefficient both with respect to photosynthesis and fluorescence.

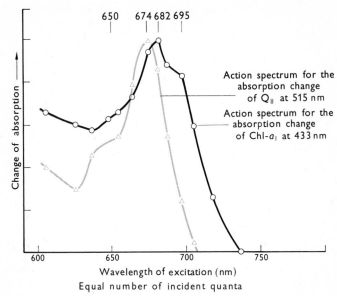

Fig. 6.7   The action spectra for the production of the absorption changes at 515 and 433 nm in spinach chloroplast fragments. The solution used for determining the action spectrum of the 515 nm change was 1.5×10^{-5} mol/dm³ dichloro-phenol-indophenol, 1×10^{-2} mol/dm³ $K_3Fe(CN)_6$, 5×10^{-2} mol/dm³, tris buffer (pH 7.2), chlorophyll concentration of 0.115 mg/cm³. The transmission at 674 nm of this suspension was 22%. The solution used for determining the action spectrum for the 433 nm change was 9.4×10^{-5} mol/dm³ PMS, 9.4×10^{-3} M sodium ascorbate, 3.8×10^{-2} mol/dm³ tris buffer (pH 7.2), chlorophyll concentration of 0.150 mg/cm³. The transmission of this suspension at 674 nm was 17%. (After Müller, A., Fork, D. C. and Witt, H. T. (1963). *Z. Naturf.*, B, **18**, 142.)

Nuclear magnetic resonance spectra

A different type of spectroscopic measurement has confirmed the possible existence of two photochemical systems. The principle of nuclear magnetic resonance spectroscopy depends on the fact that the characteristic absorption frequency of a nucleus in a magnetic field is dependent on the environment of the nucleus. Thus electron paramagnetic resonance (EPR) spectroscopy can be used to detect the formation of free radicals. A source of radiofrequency is utilized with an appropriate receiver and a device for varying a magnetic field which serves to scan the absorption spectrum.

Suspensions of algal cells and of isolated chloroplasts give a characteristic EPR signal when they are illuminated. A study of the action spectrum shows that these signals result from excitation of the far red absorbing

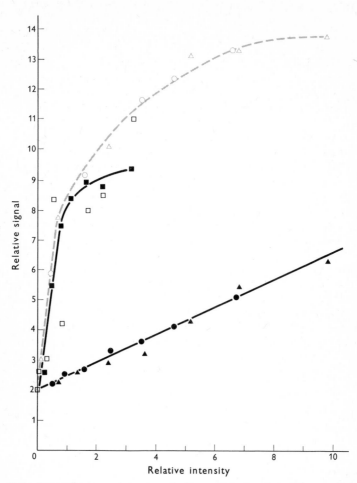

Fig. 6.8 Dependence on intensity of narrow EPR signal on light of different intensity at 713, 635 and 713 + 635 nm. Whole cells of *Anacystis nidulans* were suspended in growth medium at a chlorophyll concentration of 4.9×10^{-4}M. They were placed in a flat quartz cell of optical path 0.3 mm. The cell was introduced into the cavity and illuminated through the slotted window (temperature 20°C). Wavelength of illuminating light: squares 713 nm, circles 635 nm, triangles 635 + 713 nm. Closed symbols and full lines: data obtained with unpoisoned algae. Open symbols and upper tinted line: data obtained with a suspension containing 6×10^{-6}M DCMU. The EPR signal is produced more effectively by light at wavelength 713 nm than at 635 nm or the two wavelengths together. But in the presence of DCMU the signal is also produced by 635 nm light. This is consistent with the view that the EPR signal is produced by the oxidized form of P_{700}. (After Kok, B. and Beinert, H. (1962). *Biochem. biophys. Res. Commun.,* **9**, 349.)

forms of chlorophyll.[2] Further detailed examination in *Chlorella* has shown the signal to consist of two components; one, a signal excited by far red absorption which rapidly decreased when the light is turned off and two, a signal which persists for a longer time after darkening which can also be excited with shorter wavelengths. Kok and Beinert[83] prepared a particulate preparation from red algae from which two-thirds of the chlorophyll had been removed by acetone but most of the P_{700} remained. This preparation after dispersion by sonication showed a reversible absorption change in the light. A free radical signal was obtained of the same type as that resulting from illumination of whole chloroplasts. The signal was affected by light, ferricyanide and PMS (phenazine methosulphate) in a way correlated with changes in absorption at 700 nm. It was therefore suggested that the light-dependent short-lived EPR signal in photosynthetic material might be due to the photo-oxidized form of P_{700} (see Fig. 6.8). In *Anacystis* 713 nm light stimulated the EPR signal to a much greater extent than 635 nm light. This conclusion has been confirmed by studies of mutants of *Chlamydomonas reinhardii*. Mutant *ac-141* studied by Levine and Smillie[94,95] shows biochemical activity indicating that system II is blocked but system I is active; it also shows a fast EPR signal, but lacks the slow signal. The slow decaying signal has not yet been attributed to any particular constituent.

The biochemical significance of the two photochemical reactions is discussed further in Chapter 8.

7

The Comparative Biochemistry of Photosynthesis

Photosynthetic bacteria

The existence of photosynthetic bacteria has been known for nearly 100 years (since their discovery by Vinogradsky in 1889) but their metabolism was not carefully investigated until 30 years ago. The bacteria were known to differ in two respects from green plants. They were considered to be photosynthetic because they required light to grow; yet they did not produce oxygen, and very often they required some other chemical substance present in addition to carbon dioxide and water. They have been classified according to the pigments present (Table 7.1). Chlorophyll *a* is absent but modified forms of chlorophyll occur; bacteriochlorophyll *c* and *d* in the green bacteria and bacteriochlorophyll *a* and *b* in the purple bacteria. Bacteriochlorophyll in purple bacteria has absorption maxima at 800, 850 and 870–890 nm; in the green bacteria absorption maxima occur at 725 and 740 nm. After extraction from purple bacteria a single absorption maximum for bacteriochlorophyll at 770 nm is observed. The purple bacteria have an additional aliphatic carotenoid pigment producing the brownish-red coloration whereas the green bacteria contain carotene. In green bacteria the pigments are present in vesicles bounded by a single membrane; in purple bacteria the arrangement may be vesicular or lamellar, the latter being derived from the peripheral membrane of the cell; upon disruption the cell structure breaks up to form particles, the chromatophores.

On the basis of the additional substrate required for photosynthesis the purple bacteria have been further subdivided into two groups:

Members of the *Thiorhodaceae*, as the name implies, require sulphur present in a reduced form such as sulphide.

Members of the *Athiorhodaceae* which are primarily photo-organo-trophs require organic carbon sources for growth although some can grow on molecular hydrogen together with carbon dioxide.

Table 7.1   Classification of photosynthetic bacteria

Substrate	Group		Pigment
Sulphur e.g. H_2S	Chloraceae (or Chloro-bacteriaceae)	Green coloured bacteria which oxidize hydrogen sulphide to elementary sulphur. Strict anaerobes	Chlorobium chlorophyll: Absorption maxima 725 and 747 nm *in vivo*, 670 nm *in vitro*
Sulphur	Thiorhodaceae	Purple or brown coloured bacteria capable of oxidizing various sulphur compounds to sulphate. Certain organic substances e.g. fatty acids or H_2 gas can be used by some species	Bacteriochlorophyll Absorption maxima at 800, 850 and 870 or 890 nm *in vivo*, 770 nm *in vitro*
Simple organic compounds	Athiorho-daceae	Brown or red bacteria that preferentially utilize simple organic compounds. Some can utilize hydrogen, others inorganic sulphur compounds but growth is dependent on the presence of organic material. Many can grow aerobically in the dark	(ditto)

The green sulphur bacteria, the *Chloraceae*, use only carbon dioxide as their carbon source and inorganic hydrogen donors. They are obligatorily anaerobic, whereas the *Athiorhodaceae* possess a respiratory metabolism which can support aerobic growth in the dark, a feature which added to the confusion in thinking of them as photosynthetic organisms. All the purple sulphur bacteria can grow photosynthetically with organic substrates; certain purple non-sulphur bacteria can use thiosulphate or hydrogen gas in place of organic substrates.

In 1930 van Niel[136] studied the stoichiometry of photosynthesis in these bacteria. He showed that in certain members of the *Thiorhodaceae* sulphide became oxidized to sulphur as carbon dioxide was reduced to organic compounds and that the amount of sulphur deposited as granules of sulphur was quantitatively related to the amount of CO_2 reduced.

$$CO_2 + 2H_2S \longrightarrow CH_2O + H_2O + 2S$$

$$\frac{3CO_2 + 2S + 5H_2O \longrightarrow 3CH_2O + 2H_2SO_4}{4CO_2 + 2H_2S + 4H_2O \longrightarrow 4CH_2O + 2H_2SO_4}$$

Foster, a pupil of van Niel, showed in one species of *Athiorhodaceae* that the amount of CO_2 taken up in photosynthesis was proportional to the amount of isopropanol oxidized to acetone:

$$2CH_3 \cdot CHOH \cdot CH_3 + CO_2 \longrightarrow CH_2O + H_2O + 2CH_3COCH_3$$

Gaffron[47,48] working at about the same time but using a different organism was unable to confirm such simple quantitative relationships between carbon dioxide taken up and organic substrate used but his work was temporarily ignored.

Photosynthesis as an oxido-reduction reaction

Van Niel suggested that water in the green plant and isopropanol or H_2S in the bacteria play an equivalent biochemical role. Oxygen is the oxidation product from water in photosynthesis by green plants just as acetone is from isopropanol, and sulphur from H_2S in the bacteria. Photosynthesis was to be regarded as an oxidation-reduction reaction in which one substance accumulated in an oxidized form and an equivalent amount of CO_2 was reduced.

$$2AH_2 + CO_2 \longrightarrow CH_2O + H_2O + 2A$$

At this time it was often said that photosynthesis by green plants involved a 'splitting' or cleavage of the water molecule by light energy into a reducing [H] and an oxidizing [OH] entity; the oxidizing entity gave rise to O_2 evolution. This was a logical but not necessary corollary of van Niel's thinking.

Almost simultaneously, Hill[60] was able to show that after chloroplasts had been isolated from green leaves they were able in the light to reduce ferricyanide or quinone or indophenol dyes. His preparations were unable to react with carbon dioxide. Hill considered that in this reaction carbon dioxide had been replaced by an alternate and more easily reduced hydrogen acceptor, e.g. quinone. This interpretation was consistent with van Niel's hypothesis since when light energy 'splits' water the quinone acting as an alternative hydrogen acceptor to carbon dioxide becomes reduced to hydroquinone; oxygen is still the other product of the reaction.

Further, when certain algae were grown in the absence of oxygen, Gaffron and Rubin[49] showed that they acquired the ability to use hydrogen gas as a hydrogen donor instead of water and are capable of a type of photosynthesis called photoreduction similar to that in bacteria.

$$CO_2 + 2H_2 \longrightarrow CH_2O + H_2O$$

All these observations can be integrated into a generalized concept of photosynthesis. The reaction must have a hydrogen donating system (AH_2) and

$$2AH_2 + CO_2 \longrightarrow 2A + (CH_2O) + H_2O$$

a hydrogen accepting system (CO_2 or quinone). The light separates hydrogen from the donor and the oxidized moiety appears as one product of the reaction. The other reagent accepts the hydrogen and so accumulates in a reduced form. In the green plant the cleavage of water into a hydrogen donating system results in the dissipation of the oxidized radical as oxygen gas and the reduction of carbon dioxide to carbohydrate.

It was known at this time that photosynthetic bacteria under certain conditions evolved hydrogen in light. They could also fix atmospheric nitrogen.[75] At first these were regarded as peculiarities and not incorporated into the general mechanism of photosynthesis. But later the generalized theory suggested the possibility that the fixing of nitrogen in light represented an alternative use of the reducing power generated by the photochemical system of photosynthesis. Similarly the evolution of hydrogen gas requires the uptake of hydrogen ions from the medium and their reduction by electron donation.

The Hill reaction and ferredoxin

The essential feature discovered by Hill in the 'Hill reaction' was the capacity of isolated chloroplasts to reduce added reagents like ferricyanide, quinone and indophenol dyes using light energy:

$$2H_2O + 4[Fe(CN)_6]^{3-} \longrightarrow 4[Fe(CN)_6]^{4-} + 4H^+ + O_2$$

Even so, it was not possible to grind up all plants sufficiently and yet gently enough to preserve this activity and the observations were and still are mainly confined to preparations from leaves of spinach, sugar beet or peas. At first the substances whose reduction could be demonstrated were chemicals which did not occur naturally in leaves. The need to modify the system so that substances of biological significance could be reduced was obvious. Davenport and Hill[28] tried to recover some component which might have been dissolved out during the separation of the chloroplasts which would link the photochemical system to a wider range of hydrogen acceptors including naturally occurring substances. They found that an extract obtained from chloroplasts could link the photochemical activity of the isolated chloroplasts to reduction of cytochrome c or met-haemoglobin. They progressively purified the factor and called it met-haemoglobin reducing factor.

At the same time, San Pietro and Lang[117] sought to isolate a factor which would link the Hill reaction in chloroplasts to coenzyme reduction:

$$2H_2O + 2NADP \longrightarrow 2NADPH_2 + O_2$$

They called this photosynthetic pyridine nucleotide reductase (PPNR). At about the same time both Hill and San Pietro obtained crystalline prepara-tions. Later and quite independently other workers with the non-photo-synthetic anaerobic bacterium *Clostridium pasteurianum* isolated a factor containing iron which catalysed the fixation of nitrogen at the expense of the oxidation of pyruvate (Mortenson, Valentine and Carnahan).[102] Whatley, Tagawa and Arnon[144] showed that when the substance isolated from *Clostridium* was added to chloroplasts, it was much more effective than either the factor of San Pietro or of Davenport and Hill. All three factors when sufficiently purified proved to be the same substance, now called ferredoxin.

Ferredoxin is a protein containing non-haem iron. The crystallized com-pound was shown first by Arnon to accelerate the photoreduction of NADP by isolated spinach chloroplasts. The ferredoxin could be reduced in the dark by hydrogen gas, as an alternative to its reduction photochemically, and this involves the transfer of one electron per molecule. The reduced form was unable to catalyse the reduction of NADP directly; an additional factor shown to be a flavoprotein and subsequently named ferredoxin NADP reductase was required;[119]

$$H_2O \quad Ferredoxin \qquad Flavoprotein\ reduced \qquad NADP$$
$$\tfrac{1}{2}O_2 \quad Ferredoxin\ H_2 \qquad (NADP\ reductase) \qquad NADPH_2$$
$$\qquad\qquad\qquad\qquad\qquad Flavoprotein\ oxidized$$

it was isolated from chloroplasts and shown to have a higher affinity for NADP than NAD. Ferredoxin has now been isolated from higher plants and accounts for 1% of the non-haem iron of chloroplasts. Some properties of ferredoxin from different plant sources are shown in Table 7.2. Ferre-doxin has a standard oxygen reduction potential $E'_0 = -0.42$ volts. It shows characteristic absorption bands in the oxidized form (Fig. 7.1) but

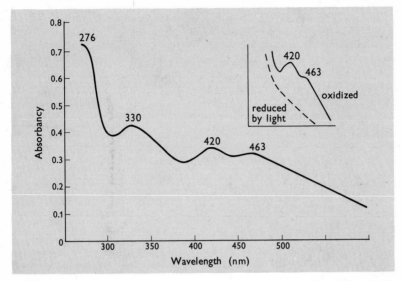

Fig. 7.1 Absorption spectrum of spinach ferredoxin. Insert shows changes in absorption at 420 and 463 nm upon photoreduction. (From Arnon, D. I. (1969). *Naturwissenschaften* **56**, 295.)

these differ slightly according to the source of the ferredoxin. Similarly the number of iron atoms per molecule varies with the source material. Ferredoxin contains a labile sulphur group which is liberated as hydrogen sulphide upon treatment with cold dilute acid or upon denaturation. There is a molar equivalence between the iron and sulphur contents and when the sulphur is removed by acidification, the iron is lost also.

Table 7.2   Properties of ferredoxins

Source	Spinach	*Clostridium*	*Chromatium*
Oxidized bands (nm)	465, 425, 325, 274	390, 300, 280	285, 300, 280
Fe/mol	2	7	7–8
S/mol	2	7	7–8
Molecular weight	12 000	6000	10 000

Ferredoxin is not specially related to the photosynthetic system; it can be reduced by dark reactions. Similarly it has the potentiality not only of reducing the flavoprotein NADP reductase but can also react with hydrogen ions to liberate hydrogen gas, or with inorganic nitrogen to form ammonia. Hence the fact that the photosynthetic bacteria do all three things is not surprising. In the photosynthetic bacteria there must be competition for the reduced ferredoxin produced by light between the carbon dioxide reducing system, the nitrogen reducing system, and in certain organisms the hydrogen evolving system. Green plants today lack the ability to evolve hydrogen or fix nitrogen gas. In the green plant the reduced ferredoxin is largely directed to CO_2 reduction although it can also be utilized in the reduction of nitrate. The photosynthetic bacteria have developed versatility at both ends of the reaction sequence; they can use a variety of hydrogen donors, whereas in the green plant water is obligatory, and at the other end they can utilize reduced ferredoxin in at least three different ways.

Photophosphorylation

Whilst the studies on electron transport in chloroplasts were progressing, Arnon et al.[6] were pursuing independent researches to see whether light energy could be used by isolated chloroplasts to synthesize ATP. They found that in the light ADP (in the presence of magnesium ions) was converted with the uptake of inorganic phosphate to ATP, provided some exogenous carrier e.g. 2-methyl-1,4-naphthoquinone (vitamin K_3) was added. A surprising result was that the only product of the light reaction was ATP without any accompanying reduced product so that the system had changed from a system generating reducing power to one producing phosphate bond energy. This process in which no product appeared other than the phosphorylated compound was called cyclic photophosphorylation since any electron flow involved must be in a closed cycle.

$$ADP + P_i \xrightarrow{\text{light}} ATP + H_2O$$

A whole range of substances could catalyse this process provided that they could be reversibly oxidized and reduced and had a standard electrochemical potential not very far from zero. One of the most effective is methylphenazonium methyl sulphate (PMS) which in its reduced form becomes pyocyanine.[62] A similar reaction with chromatophores isolated from a photosynthetic bacterium was shown by Frenkel.[45]

If these special carriers were omitted from the reaction mixtures but one of the conventional Hill reagents, e.g. potassium ferricyanide, was added a limited photophosphorylation was observed. Furthermore, the addition of ADP, P_i and magnesium ions caused an acceleration of the rate of reduction of the ferricyanide. A stoichiometrical relationship between the amount of ferricyanide reduced (two moles) and the amount of ATP formed (one mole) was claimed, i.e. the processes were obligatorily coupled, but the conflicting results obtained by later workers show the difficulty of quantitatively relating electron flow to phosphorylation. Values varying between one and four molecules of phosphate taken up per two electrons transferred have been reported. The process was called non-cyclic phosphorylation because it involved a second reagent which became reduced:

$$H_2O + NADP + ADP + P_i \longrightarrow NADPH_2 + ATP + \tfrac{1}{2}O_2$$

One important difference between cyclic and non-cyclic phosphorylation was the relationship between the wavelength of the exciting light and the rate of reaction. In the critical region around 700 nm cyclic photophosphorylation increased in rate for a given amount of light energy as the wavelength was increased, but the non-cyclic system showed a decrease in rate with increasing wavelengths just as for photosynthesis *in vivo*.[118] Another important difference between cyclic and non-cyclic photophosphorylation is shown by the effect of certain inhibitors. Substituted ureas, e.g. DCMU, do not inhibit the cyclic system catalysed by phenazine methosulphate, whereas the non-cyclic system in which ferricyanide is reduced is inhibited.[70] These facts suggest that cyclic photophosphorylation is dependent only on photochemical system I. Certain agents, e.g. ammonium sulphate, which do not inhibit the oxidation-reduction reaction but inhibit phosphorylation in the non-cyclic process (called uncoupling agents) affect the two types of photophosphorylation to different extents.[7]

There is some evidence that cyclic phosphorylation occurs *in vivo* and is not an artefact produced *in vitro* after addition of special reagents. The evidence is indirect and based on the assumption that, when a plant is transferred from air to nitrogen, oxidative phosphorylation in the mitochondria will be decreased to a significant extent (since the reactions generating ATP through oxidation reactions are suppressed), but that the generation of ATP photosynthetically should continue unchanged. On this basis illumination in nitrogen should permit those reactions which involve ATP utilization. For example, Maclachlan and Porter[96] showed that disks cut from tobacco leaves and placed in sugar solutions in oxygen formed starch, but in the presence of nitrogen in the dark, little starch was synthesized. Yet if the tissue was illuminated even more starch was formed than in oxygen in the dark. Thus a reaction known to require phosphorylated nucleotides was shown capable of being driven by light under anaerobic conditions.

Another example demonstrated by Kandler and Tanner[77] concerns the assimilation of glucose by *Chlorella* cells. *Chlorella* accumulates sugar from a weak solution by a process dependent in the dark on aerobic metabolism. The process is suppressed when the cells are put into nitrogen, but is restored by light. It was shown further that the effective wavelengths of light extended into the far red region of the spectrum; also the process was relatively insensitive to the addition of DCMU.

A similar effect upon the uptake of potassium ions by *Nitella* was shown by MacRobbie.[97] The movement of ions into the cell is an active process, normally requiring oxygen. In addition it could be stimulated by light in the presence of nitrogen. Again inhibitor evidence suggested that only cyclic photosynthetic phosphorylation was involved. In *Chlorella* the formation of an adaptive enzyme, isocitric lyase, has been shown by Syrett[128] to depend on aerobic metabolism. In nitrogen the process continues only if the cells are illuminated. All these examples provide indirect evidence that high energy phosphate has been generated by light in nitrogen. Since illumination is effective even for wavelengths longer than 700 nm and the effects are insensitive to inhibitors such as DCMU, the examples indicate the probability of cyclic photosynthetic phosphorylation *in vivo*.

The role of light energy in photosynthetic bacteria

The study by Gaffron with a member of the *Athiorhodaceae* referred to earlier in this chapter showed that acetate metabolized in the light anaerobically gave rise to a product more reduced than carbohydrate. Gaffron had regarded the reaction as a light-catalysed assimilation of organic substance, in contradiction to van Niel's view which considered that the organic substance acted solely as a hydrogen donor. Later in 1959 the product was identified as poly-β-hydroxybutyric acid $(C_4H_6O_2)_n$.[123] The polymerization involves the intermediate formation of β-hydroxy-butyryl Co A and the formation of this requires the utilization of one molecule of ATP for each molecule of acid utilized.

$$2 \text{ Acetate} + 2\text{Co A SH} + 2\text{ATP} \longrightarrow 2 \text{ acetyl Co A} + 2\text{AMP} + 2\text{P}_i$$
$$2 \text{ acetyl Co A} \longrightarrow \text{acetoacetyl Co A} + \text{Co A SH}$$
$$\text{acetoacetyl Co A} + \text{PNH}_2 \longrightarrow \beta\text{-OH butyryl Co A} + \text{PN}$$

It was shown that in light the presence of hydrogen gas resulted in a more rapid conversion of acetate to the polymer; it may be assumed that the hydrogen reduces pyridine nucleotide in the dark and that the only need for light energy is to catalyse the formation of ATP which is used for the formation of acetyl Co A. If however the environment is less reducing, light energy may be also required to provide reducing power as well as phosphorylated bonds. In that case reduction of acetoacetyl Co A will compete for reduced coenzyme with the reduction of carbon dioxide. Hence according to the conditions of the experiment, the products of photosynthesis will vary and the bacteria have a biochemical plasticity with respect to the utilization of light energy. For example, in *Chromatium*

when acetate is fed in the light in the presence of helium, only 7% is converted to poly-β-hydroxy butyrate, whereas in the presence of hydrogen
gas 74% of the acetate is assimilated as polymer.

The view is sometimes taken that the essential function of bacterial
photosynthesis is the formation of ATP. This requires that where electron
transfer reactions also take place in the process it would be necessary for
the accumulated ATP to be used to drive electron flow giving a reversed
electron transport at the expense of the energy of ATP (or of some other
high energy intermediate which under other circumstances can lead to the
formation of ATP).

8

Electron Transport in Photosynthesis

Cytochromes and photosynthesis

After the demonstration of photophosphorylation by isolated chloro-plasts, the possible function of cytochromes in photosynthesis was inten-sively investigated. The non-green parts of plants contain members of c and b types of cytochrome characterized by their redox potential and absorption bands in the reduced form. In the green parts of plants a cyto-chrome of the c class, cytochrome f (from the Latin *frons*, a leaf) is present which differs slightly both in its higher potential ($E'_0 = +0.365$ V) and in its absorption maximum (α band at 555 nm) from normal c. Non-green tissues contain cytochromes of the b type, b_3 and b_7 with an oxidation-reduction potential near zero volts, and with characteristic absorption spectra. The corresponding cytochrome in the photosynthetic tissue, b_6 or cytochrome b_{563} (α peak 563 nm), has a standard potential $E'_0 = -0.04$ V.[10] Both cytochromes f and b_6 are bound to the chlorophyll-containing thyla-koid membranes of the chloroplast. Cytochromes are also present in the photosynthetic bacteria;[65] they are not simply classified but in general the bacteria lack low potential b type components.

In respiration the electron chain transfers electrons from coenzyme to cytochrome b, then from b to c, and from c to oxygen through cytochrome a_3. The essential feature of this process is that the fall of electrochemical potential is divided into convenient steps and the energy liberated at each step utilized to convert ADP to ATP. The cytochrome chain transfers some of the energy of oxidation into phosphorylation and it is the passage of electrons from a more reducing cytochrome to a more easily reduced cytochrome which plays an essential part.

If the two types of cytochrome characteristic of higher plant and algal

photosynthetic tissue are to be used in a way similar to the role of cytochromes in respiratory metabolism it would be necessary for cytochrome
b to react exothermally with cytochrome f, so liberating energy to permit
the formation of ATP. This reaction step was an essential feature of the
mechanism of photosynthesis proposed by Hill and Bendall.[61] If light
energy were used to reduce a b type cytochrome this could in turn be used
to reduce a c type cytochrome, cytochrome f, and that process could be
accompanied by phosphorylation. If nothing else happened this would
constitute a cyclic photophosphorylation system. Light energy is required
to reduce cytochrome b_6 either directly or indirectly which then in turn
reduces cytochrome f. Since it is known that in order to operate the cyclic
system, a redox substance with a standard potential in the range between
0.0 and -0.4 V must be added, it follows that the added catalyst acts as a
link between ferredoxin reduced by the light reaction and cytochrome b_6
(see Fig. 8.1). The added catalyst is necessary to compete with the flavo-

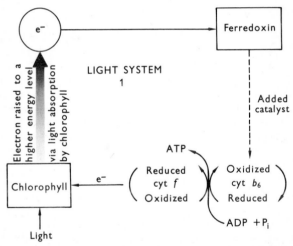

Fig. 8.1   Cyclic photophosphorylation. Chlorophyll absorbs a photon of light of
sufficient energy; this sends an electron into a high energy state. The electron
reduces ferredoxin and in turn the cytochrome system via the added carrier. The
reduction of cytochrome is coupled with the phosphorylation of ADP to form
ATP. The electron returns, at a lower energy state, to the chlorophyll which had
acquired a positive charge after the initial ejection of the electron. No 'external'
electron donor is necessary for this process.

protein with which the ferredoxin alternatively reacts prior to reduction of
NADP. Since cyclic photophosphorylation is photosensitized by wavelengths longer than 690 nm the reaction requires photosensitization only

by system I pigment. Therefore this photosystem must photocatalyse the reduction of ferredoxin and the oxidation of cytochrome f.

In the photosynthetic bacteria certain of the exogenously added redutants, e.g. H_2 gas, are capable of reducing ferredoxin exothermally. Light energy may still then be required for photosynthesis but presumably only to catalyse photophosphorylation resulting from electron flow through the cytochromes. In bacteria using a less powerful reducing substrate, e.g. H_2S, light energy may be necessary to energize electrons for the reduction of ferredoxin from the oxidation of the exogenous substrate. But even then the reduction may be brought about by reverse electron flow consequent upon the formation of ATP (as mentioned in Chapter 7). The bacteria use NADH (as opposed to NADPH in higher plants) but attempts to demonstrate quantitatively the reduction of NAD, or of any other low potential reductant, by light have failed. In *Chromatium* two cytochrome components, cytochrome 552 and cytochrome 555, have been observed and the action spectra for their oxidation differ. Probably cytochrome 555, $E'_0 \simeq 340\,mV$, is involved with a cyclic electron flow resulting in phosphorylation whereas cytochrome 552, $E'_0 \simeq 10\,mV$, may be concerned with electron flow from the exogenous substrate. But whether it is justified to consider that there are two photochemical systems in any bacterium is doubtful.[50]

The special feature of green plants is the ability to utilize water to reduce ferredoxin, a reaction which is strongly endothermic. This according to Hill and Bendall might be achieved in two photochemical steps using the cytochromes as a half-way house toward the reduction of ferredoxin. Another photochemical system, system 2, using a different wavelength of light was postulated as necessary to permit electrons from water to reduce

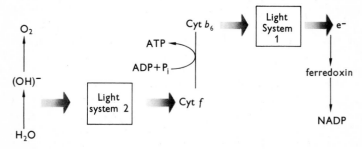

Fig. 8.2 The role of the cytochromes as intermediates between the two light reactions in green plants according to the suggestions of Hill and Bendall.[61]

cytochromes (see Fig. 8.2). Thus system II light was used to effect the reduction of cytochrome f by water; this could then react thermally with cytochrome b_6 which in turn, after excitation with system I light, could reduce ferredoxin. By passing through the two cytochromes, but in a direction from *more* reducing to *less* reducing, phosphorylation could take place between the two light steps. For each electron transferred to ferredoxin and each half atom of oxygen produced from water, one molecule of ADP will be formed, producing an obligatory stoichiometric relationship between the oxido-reduction and the coupled phosphorylation reactions. Hence, the operation of both systems I and II results in a non-cyclic type of phosphorylation whereas the operation of system I (with an additional added carrier) suffices for cyclic phosphorylation.

Arnon[5] proposed an alternative hypothesis. Whilst he accepted the physiological evidence that two photochemical systems were involved in photosynthesis, he did not accept Hill's proposal as to their biochemical significance. He considered that system I, absorbing far red light, was concerned exclusively with cyclic photophosphorylation; system II alone was considered to effect the transfer of electrons from water to ferredoxin. During the latter process non-cyclic phosphorylation took place so that Arnon's mechanism proposed two distinct reaction sequences producing photophosphorylation, whereas Hill and Bendall's scheme had the simplicity that the same reaction step catalysed phosphorylation under cyclic and non-cyclic conditions. Arnon was mainly influenced by the evidence that it was difficult to demonstrate an enhancement effect for the reduction of NADP by water in isolated chloroplasts. It is still not clear under what conditions such an effect can be observed, since some workers claim to have observed it whilst others have failed.

In a later paper Knaff and Arnon[80] modified this proposal, since they recognized that longer wavelengths were effective in the reduction of NADP using reduced dichlorphenol indophenol as electron donor in chloroplasts but not when water was the electron donor. They therefore proposed that the process of electron transfer in chloroplasts should be separated into two systems, described as photosystem IIa and photosystem IIb linked by another cytochrome, cytochrome b_{559} and plastocyanin. They still considered cyclic photophosphorylation to be activated by a completely separate photochemical system, system I, involving cytochromes f and b_{563}. There seems no convincing evidence which would lead one to prefer the hypothesis of Arnon to the simpler view proposed by Hill and Bendall; neither does it explain the antagonistic effect of red and far red light in oxido-reduction of cytochrome f.

The separation of two photochemical systems

Hill and Bendall chose to regard reagents such as reduced indophenol as introducing electrons part way along the sequence of reactions between systems I and II. Thus by using suitable reagents it is possible to demonstrate the reactions of system I separated from those of system I and II combined. For example, reduced indophenol can donate electrons to cytochrome f. If therefore chloroplasts are supplied with reduced indophenol (achieved by adding indophenol together with excess ascorbate) ferredoxin can be reduced when chloroplasts are illuminated with light absorbed by system I only.[137] Also, if a chloroplast preparation is allowed to age, it is found that the reaction using reduced indophenol persists for a longer period than that using water as the electron donor; consistent with this, cyclic photophosphorylation is less affected by the ageing process than straight chain phosphorylation. Again, by the use of the selective poison DCMU, the electron chain can be interrupted between system I and system II. In the presence of the poison, system I can still be demonstrated as active, e.g. by using reduced indophenol as electron donor. It is still not clearly established whether it is possible to isolate system II by adding a suitable hydrogen acceptor which can accept electrons immediately from it, but there is evidence suggesting that under certain conditions indophenol and ferricyanide can act in this way. For example, with well washed chloroplast preparations some of the soluble carriers which link system I and II can be washed out and the role of electron acceptors for system II most easily demonstrated.[84] Thus by the choice of suitable reagents or by the differential effect of poisons or ageing, the properties of system I and of systems I and II can be separately studied in chloroplast preparations. The absorption for system I alone is best determined as the action spectrum for cytochrome f oxidation in the presence of DCMU.

Izawa and Good[68] titrated the minimum number of molecules of DCMU which could completely inhibit the reduction of NADP by water catalysed by isolated chloroplasts. It was found that one molecule of DCMU for each 2000 molecules of chlorophyll present was just sufficient to suppress the reaction completely. Therefore they suggested that there must be some component present in system II in this relatively limited quantity; it has not yet been identified chemically.

Carriers in electron transport

Other natural carriers which may play a part in the electron flow of photosynthesis have been isolated from chloroplasts. In 1960 Katoh[78] isolated from *Chlorella* and from leaves of higher plants a copper-containing protein, plastocyanin (mol. wt. 21 000), in an abundance of approximately 1 mole for each 400 moles chlorophyll; it contains two copper atoms per molecule. After treatment with detergents chloroplasts were no longer able to reduce coenzymes using water; but if plastocyanin was then added the ability was restored. The same result was obtained when the reaction tested was system 1 alone, i.e. between reduced indophenol and coenzyme or cyclic photophosphorylation. Hence plastocyanin is a component which must be in that part of the electron path common to systems I and II. The standard potential of plastocyanin is $+0.40$ V and it is therefore presumed to react at a point near to cytochrome f. The reduced form is colourless and not oxidized by air; the oxidized blue form (absorption bands at 460, 597 and 770 nm) is rapidly reduced by chloroplast preparations in the light. The two components, plastocyanin and cytochrome f, are present in about the same abundance. Some authors consider that there may be two parallel electron paths, some electrons reacting through cytochrome f and others through plastocyanin. Selected mutants of *Chlamydomonas* were studied by Levine,[93] some of which were devoid of cytochrome 553 (*ac-206*) and others devoid of plastocyanin (*ac-208*). The absence of plastocyanin reduced cyclic electron flow but the loss of cytochrome 553 had little effect, suggesting that cytochrome 553 may serve as an additional reservoir for the storage of electrons but is not an essential constituent of the reaction path.

Mutants of the green algae *Chlamydomonas reinhardii*, *Euglena gracilis* and *Scenedesmus obliquus* have been obtained in large numbers by Levine. Each mutant strain has a reduced rate of NADP reduction with water as the electron donor and this can be correlated with the presence or absence of certain electron carriers. A mutant of *C. reinhardii ac-80$_a$* lacks P_{700} as shown by the loss of its characteristic difference spectrum; it can carry out the Hill reaction but only with DPIP and not with NADP as acceptor. Also the light-induced oxidation of cytochrome 559 and cytochrome 553 is not observed in this organism. Cyclic phosphorylation catalysed by PMS also cannot be demonstrated (see Fig. 8.3).

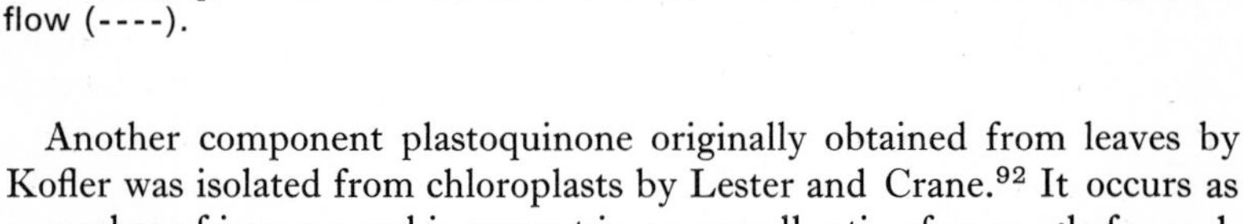

Fig. 8.3 Electron transport chain in *C. reinhardii* according to Levine with sites indicated (∖) where mutant blocks have been observed and also showing cyclic flow (----).

Another component plastoquinone originally obtained from leaves by Kofler was isolated from chloroplasts by Lester and Crane.[92] It occurs as a number of isomers and is present in an overall ratio of one mole for each 7 moles chlorophyll. The structure is similar to that of ubiquinones. It has been shown to be oxidized and reduced in the course of photosynthesis, reduction being accompanied by a decrease in optical density at

255 nm; excess of system I light tends to oxidize it and excess of system II to reduce it. The photoreduction is inhibited in the presence of DCMU. The standard oxidation-reduction potentials lie between $+0.00$ V and $+0.10$ V and plastoquinones are assumed to operate at the other end of the thermal reaction step from plastocyanin (see Fig. 8.4). Extraction of

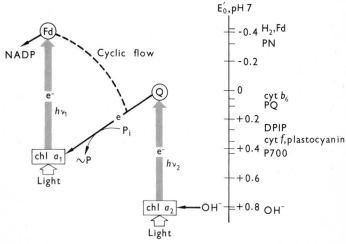

Fig. 8.4 Generalized diagram of two light reactions in photosynthesis. The oxidation-reduction potentials (E'_0, standard, pH 7) of some possible carriers is given on the right. Fd, ferredoxin; PN, pyridine nucleotide; PQ, plastoquinone; cyt, cytochromes; P_{700}. Q is the fluorescence quencher.

plastoquinone (with light petroleum) has rather less effect on cyclic phosphorylation than on non-cyclic photophosphorylation in chloroplasts, presumably because electrons from cyclic flow must react between plastoquinone and a cytochrome.

A b type cytochrome isolated from chloroplasts, cytochrome 559, is another candidate to fill the position assigned in the discussion earlier in this chapter to cytochrome b_6. It is believed to exist in two distinct forms which may differ in potential by about 300 mV (i.e. $+0.065$ V and $+0.37$ V); the low potential form may play a role in cyclic electron flow. Indeed some consider that cytochrome f is the only cytochrome which operates in the electron flow between systems I and II; cytochrome 559 is then considered to be associated in a side reaction.

P_{700} is the component present in chloroplasts responsible for the marked change in the absorption about 700 nm during illumination. By the addition of reagents of known potential and observing the effect on the absorption at 700 nm, the standard oxidation-reduction potential of P_{700} has been found to be $E'_0 = +0.4$ V. Hence P_{700} has been considered to be the photocatalyst which sensitizes system I (see Fig. 8.5).

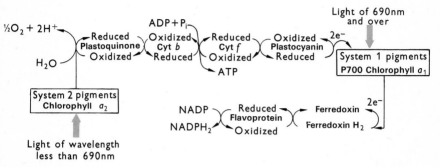

Fig. 8.5   A scheme to show the carriers concerned in electron transport between the two photochemical system of photosynthesis.

Ferredoxin has been frequently proposed as the primary electron acceptor for photosystem I. Suggestions have also been made that other more reducing compounds act as primary acceptor, with potentials as low as -0.6 V, ferredoxin then being secondary. Fuller and Nugent[46] proposed pteridines, Yocum and San Pietro[148] a ferredoxin reducing substance (FRS), others a flavin or a photosynthetic pigment identified by its absorption change at 430 nm (P_{430}).

A small part of the bacteriochlorophyll at the reaction centre in photosynthetic bacteria is considered to be in a special form, called P_{890} or P_{870} according to the wavelength at which maximum change in absorption takes place upon illumination. Most of the bacteriochlorophyll can be destroyed without affecting the absorption change at 890 nm and probably only 1% of the total pigment is in the reaction centre. The absorption change is due to oxidation since it can also be brought about by addition of ferricyanide. Other changes in absorption result from the cytochrome components, probably of the c types and a ubiquinone; these constituents form the reaction centre of bacteria. Possibly a second cytochrome of the b type is also present. Transfer of electrons from pigment through ubiquinone and cytochrome forms a cyclic reaction path which can convert light energy to phosphate energy, a necessary requirement for all photosynthetic organisms (Chapter 7). There is the possibility of a second light reaction coupled to the oxidation of substrate and an additional cytochrome component; this

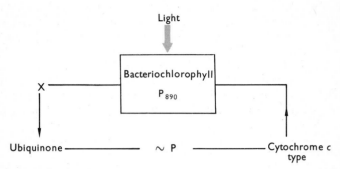

Fig. 8.6   Cyclic electron transport in bacteria.

is also dependent on light excitation by bacteriochlorophyll (see Fig. 8.6) but its existence is still disputed.

Physical separation of two particles

The possibility exists that the two photosystems in green plants occur as physically discrete entities within the chloroplast *in vivo*. To try to separate them chloroplasts have been treated with various detergents like digitonin or Triton × 100 or other dispersing agents. Using digitonin followed by differential centrifugation Anderson and Boardman[4] obtained a less heavy and a heavy particle fraction. The first striking difference between them was the ratio of chlorophyll *a* to chlorophyll *b*; the light system had a ratio of anything between 4 and 7, the heavier system a ratio 2, whereas the ratio for the whole tissue was about 3/1 (see Table 8.1). Hence one

Table 8.1   Composition of particles obtained by Boardman and Anderson after treating chloroplast suspensions with 2–4% detergent digitonin and separating two sizes of particles by differential centrifugation.

	Whole tissue	Smaller particles	Heavier particles
cyt f/chlorophyll	1/400	1/345	1/730
P_{700}/chlorophyll	1/420	1/200	—
cyt f/cyt b_6	1/3.6	1/2	cyt 559 present
chl a/chl b	3/1	5.7/1	2/1

fraction was relatively enriched in chlorophyll *a* and the other in chlorophyll *b*. The ratio of cytochrome *f* to the total chlorophyll present for the whole tissue is normally about 1/350 but the light fraction was found to have relatively more. On the contrary, with respect to cytochrome b_6, relatively more was in the heavy fraction. Cytochrome b_{559} is unequally distributed between the particles. These results suggest that the light particles are relatively enriched with respect to system I and the heavy with respect to system II, and are consistent with the view that a physical separation of two particles containing the separate photochemical systems has been achieved. The ability to utilize reduced indophenol to reduce coenzyme photochemically is shown equally by the heavy particles and the light but the heavy particles more actively catalyse reactions believed to require photochemical system II. It is still not proven whether the two types of particle actually exist in the chloroplast before treatment, or whether the detergent has broken up the system of membranes into two distinct forms. One possibility is that detergents selectively destroy certain components. The general evidence favours the view that separate particles occur *in vivo* and are liberated by suitable detergent treatment with their biochemical integrity preserved. Other methods of preparation have included sonication treatment[69] and exposure of chloroplasts to very high pressures;[99] by means of these various methods the same two types of particle can be prepared.

Mechanism of phosphorylation

Phosphorylation was first considered in relation to a reaction between an oxido-reduction pair, say AH_2 and B. The reduced product BH_2 was believed to react with an intermediate I and then a hydrogen acceptor C; part of the energy of the oxidation reaction was transferred to the linkage between I and B, $B \sim I$.

$$AH_2 + B \longrightarrow BH_2 + A$$
$$BH_2 + I \longrightarrow BH_2\text{-}I$$
$$BH_2\text{-}I + C \longrightarrow CH_2 + B \sim I$$
$$B \sim I + ADP + P \longrightarrow I + B + ATP$$

The 'high energy compound' $B \sim I$ could then react with ADP and phosphate to give ATP. Such a mechanism postulates that there must be a high energy compound formed in the system prior to phosphorylation. Hind and Jagendorf[63] sought to demonstrate such a high energy intermediate by illuminating chloroplasts in the absence of phosphorylating reagents but in the presence of a suitable carrier (e.g. PMS) to permit electron flow.

When phosphorylation reagents were then added in the dark, phosphorylation took place, thus separating in time the formation of a high energy complex in the light from its subsequent use in phosphorylation in the dark. They were able to show that the ability generated by light in the absence of phosphate subsequently to effect phosphorylation persisted for some seconds in the dark, but they were unable to isolate a specific intermediate. An inhibitor of electron flow (e.g. DCMU) was inhibitory if added in the light but not if added only in the dark.

An alternative mechanism of phosphorylation suggested by Mitchell[100] proposed that the trapped light energy was not within a chemical compound but present as a physical state capable of generating electromotive force. For example, if during illumination an electrical potential difference is established between the inside and outside of the chloroplast, this stored electrical energy could subsequently be utilized for phosphorylation. Mitchell supposed that the chloroplast membrane was such that during illumination hydrogen ions moved selectively across it to establish a difference in hydrogen ion concentration between the two sides. The synthesis of ATP ions was brought about by the reversal of hydrolysis, the ATP-ase of the membrane being activated by the flow of hydrogen ions across the membrane until equality was again reached. It follows that chloroplasts should accumulate hydrogen ions as a result of illumination in the absence of phosphorylation reagents. Hind and Jagendorf[63] were indeed able to demonstrate that light did induce pH changes in a suspension of chloroplasts in the absence of phosphorylating agents, the medium becoming alkaline. It may not be a coincidence that a sequence of intermediates in the electron transport chain of chloroplasts can be arranged as alternate hydrogen and electron acceptors. Figure 8.7 shows how a suitable spatial arrangement of these intermediates could operate in the manner envisaged by Mitchell so that the passage of electrons through carriers in a membrane is accompanied by the establishment of a proton gradient across the membrane.

There are several important differences between the chemical hypothesis and the mechanism proposed by Mitchell. The chemical hypothesis requires the presence of a specific uncoupling protein, presumably at a localized site within the particular membrane. The Mitchell hypothesis requires an asymmetric membrane and the energy to be accumulated as a difference in ion concentration across it. Certain chemicals such as ammonium sulphate have been shown to inhibit photophosphorylation whilst permitting electron transfer to continue in chloroplasts. According to the chemical view, such uncouplers must preferentially react with the high energy intermediate preventing its reaction with ADP; according to the other view such agents must modify the membrane so that it is no longer able to maintain a difference in ion concentration. On either view it follows

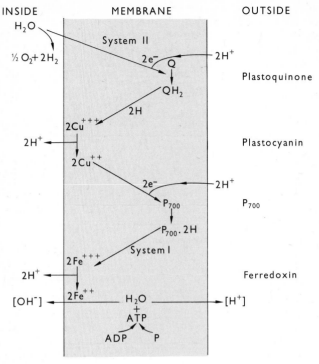

Fig. 8.7   The sequential reaction of electron carriers in the chloroplast giving rise to hydrogen ion movements according to the theory of Mitchell.

that reaction with ADP involves a rate limiting reaction because it has been shown that when uncoupling agents produce a decrease in the rate of phosphorylation there is a concomitant increase in the rate of electron transfer.

Conformational changes in chloroplasts

Other methods have been used to study the formation of a high energy state. For example, the average size of chloroplast particles has been shown to increase during illumination.[109] Under other conditions shrinkage has been observed (see Fig. 8.8). The type of response is determined by the nature of the anions present. Shrinkage is reversible in the dark but swelling is not. Such effects indicate changes in physical state. Also by studying the optical properties of a chloroplast suspension it has been shown that light is scattered to a different extent when the suspension is

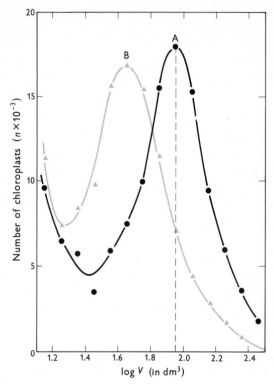

Fig. 8.8 The volume-distribution curves of whole chloroplasts in the dark (curve A) and in the light (curve B). (From Itoh, M., Izawa, S. and Shibata, K. (1963). *Biochim. Biophys. Acta*, **66**, 319.)

illuminated. These physical changes are assumed to be correlated with the formation of a high energy state. By treating chloroplasts with a chelating agent such as EDTA, a specific factor, the coupling factor, can be extracted. After such treatment chloroplasts still show changes in pH and in light scattering and in swelling upon illumination, but they cannot catalyse phosphorylation. When the coupling factor is added back, phosphorylation is restored. Together with the change in pH, movement of other ions in and out of the chloroplast, e.g. potassium, has been observed upon illumination.[31] These effects are not yet fully investigated.

Acid–base transfer and phosphorylation

According to the hypothesis of Mitchell it follows that generation within the chloroplast of an electrochemical potential difference by any means other than light should result in a state potentially capable of generating ATP. In agreement with this prediction Hind and Jagendorf[64] found that if chloroplasts were equilibrated at an acid pH 4.0 in the dark with ADP and inorganic phosphate and the pH was then rapidly raised to 8.0, phosphorylation resulted. Moreover, the maximum synthesis of ATP observed was stoichiometrically related to the number of protons passing through the membrane during the equilibration of hydrogen ions. Later it was confirmed that a change in pH was more important than the absolute value of the initial or final pH. The production of ATP can be so large (up to 1 mole ATP/4 moles chlorophyll) as to make it extremely unlikely that the phosphorylation can be due to the presence of some chemical intermediate whose state is changed by the change in pH. The phosphorylation resulting from an acid–base transfer in the dark is specifically inhibited by a serum prepared from the coupling factor mentioned previously. The dark phosphorylation is also uncoupled by these agents believed to act as proton conducting reagents, e.g. nitrophenols or carbonyl cyanide m-chlorophenyl-hydrazide (CCC).

When spinach chloroplasts are rapidly changed from an acidic to a basic suspension medium, Mayne and Clayton[98] observed that they emit for a brief time chlorophyll fluorescence. It is considered that chlorophyll has been excited to the singlet state at the expense of some high-energy state, a phenomenon demonstrated by Strehler and Arnold[127] with algae some years earlier.

The photosynthetic reaction centres

The abundance of each intermediate carrier relative to chlorophyll has been determined. One molecule of cytochrome f is present for each 350 molecules of chlorophyll a. Plastocyanin is in about the same proportion as cytochrome f and the proportion of P_{700} is also the same. Hence a system I particle may be visualized as containing one plastocyanin, one cytochrome f, one P_{700} and something like 300 chlorophyll a molecules. The absorption of light can take place in any one of a number of molecules. The energy is transferred to a single molecule (e.g. P_{700}) at the reaction centre at which electron transfer first takes place. Energy 'harvesting' by the pigment 'antennae' is one way of describing the absorption by the bulk of the pigment molecules which are not directly involved in the initiation of the chemical process. Again, there are relatively twice as many cytochrome b_6 molecules as cytochrome f, and there are three molecules of plastoquinone for each molecule of b_6. Hence for each molecule of plastoquinone there are approximately 50 chlorophyll molecules. Thus again several chlorophyll molecules must be associated with one centre of reaction for photochemical system II. In 1932 Emerson and Arnold[38] had postulated a photosynthetic unit on the basis of physiological studies with flashing light upon photosynthesis in algal cells. The exciting illumination was sufficiently intense to excite every chlorophyll molecule, but with a sufficiently brief flash no molecule could be excited twice within the duration of a single flash. The average maximum photosynthetic yield obtained from such an intense but very brief flash is a measure of the absolute quantity of some yield-limiting intermediate between the chlorophyll and the ultimate biochemical product, in this case measured as oxygen. With a flash of 10^{-6} s duration so bright that further increase in energy did not increase the photosynthetic yield, Emerson and Arnold found a limiting maximum yield per flash of 1 molecule of oxygen for every 2000 moles of chlorophyll; they measured the production of oxygen with a manometer so that the yield per flash could be obtained only as an average from a sequence of several hundred flashes. Hence within the sequence of intermediates which together form the electron chain, there must be one component present at most in this relative quantity. A similar limiting factor was indicated from the results of experiments with DCMU referred to earlier in this chapter.

Thus in present terms we can visualize two types of particle each with approximately 300 to 400 molecules of chlorophyll and with certain specialized carriers which differ in the two particles and with a particular molecule at the reaction centre of each which participates directly in the primary chemical reaction. This is probably P_{700} in system I but has not yet been identified for system II. Such entities may one day be identified using the electron microscope. At the present it is still doubtful which features of chloroplast structure observed in the microscope are essential for photosynthesis. For example, attempts to correlate the extent of lamellar stacking to form grana with the ability to evolve oxygen have not been convincing.

In recent studies Joliot[74] has measured the oxygen produced from a single light flash with chloroplasts or algae using a sensitive polarographic method. The flashes were intense but short (10^{-5} s) and separated by dark intervals of seconds. If such a sequence of flashes follows a prolonged dark period, the first flash, no matter how intense, produces no oxygen, the second very little, the third more and the fourth approximately the same as the average of many subsequent flashes. The same relative magnitudes are observed with flashes 5–8 but with smaller amplitude and so on in subsequent cycles. Since it is known that four electron transfers are required for the release of one molecule of O_2 it is proposed that the accumulation of individual changes takes place in the first two flashes and in the third and fourth these may be used in units of four. The experiments clearly indicate that the reaction centres of system II must act independently in the accumulation of the four separate electrons necessary for the release of one molecule of oxygen from water.

Appendix

Lambert's Laws are concerned with the amount of light passing through an absorbing material as a function of the thickness and of the light entering the material.

Beer's Law specifies the effect of the concentration of an absorbing substance on the absorption of light by a solution.

The fundamental law which embodies the conclusions of both Beer and Lambert is sometimes called the ***Beer–Lambert Law.*** It is:

$$\log_{10}(I_0/I) = ktc \tag{1}$$

where I_0 = intensity of incident light,
$\quad I$ = intensity of light after passage through the absorbing layer,
$\quad t$ = thickness of the absorbing layer,
$\quad c$ = concentration of solution forming the absorbing layer,
$\quad k$ = a constant, which is characteristic for the absorbing substance.

A number of definitions arise from this law:

1. ***Absorbance*** = D (also called Density, Optical Density, Extinction, Absorbancy)

This is the measured value $\log I_0/I$

$$\text{i.e. } \log \frac{I_0}{I} = D$$

2. ***Molar extinction coefficient***

Molar Extinction Coefficient, R(or ε) = $\log I_0/I$ for a 1M solution in a cell with a light path of 1 cm.

Thus in equation (1) where t is measured in cm and c in molarity, k (sometimes denoted as ε) = Molar Extinction Coefficient.

3. **Specific extinction coefficient** = $\varepsilon'1\%/1$ cm = $\log I_0/I$ for a 1 cm depth of a 1 per cent w/v solution.

Example of the use of Extinction Coefficients:

A solution of ATP at pH 7.0 has an absorbance of 0.75 when measured in a cell of 1 cm light path at a wavelength of 259 nm.

The Molar Extinction Coefficient for ATP under these conditions is $15\,400/\mu\text{mole/cm}^3$. What is the concentration of the ATP solution?

A 1 μM-solution would have an extinction of 15 400.

An extinction of 0.75 corresponds to a molarity of

$$\frac{0.75}{15\,400} \times 1 = 4.87 \times 10^{-5} \text{ M} = 0.0000487 \text{ M} = 0.0487 \text{ }\mu\text{moles/cm}^3$$

Measurements in Absorption Spectrophotometry may be made also in terms of **Per Cent Transmission.**

$$\text{T} = \text{Per Cent Transmission} = I/I_0 \times 100$$

i.e. this is simply the percentage of light entering an absorbing layer which emerges at the other side.

Per Cent Transmission is especially useful in considering the amount of energy transmitted by a filter with a light source of known spectral energy distribution (i.e. for which the amount of energy emitted at various wavelengths is known).

Other definitions of importance:

1. Units characterizing Light.
   (a) **Wavelength** (λ) is commonly expressed in the following units:

   1 μm (1 micrometre or micron) = 1×10^{-6} metre, i.e. one millionth of a metre.

   1 nm (1 nanometre)   = 1×10^{-9} metre, i.e. one thousandth of a micrometre (originally 1 mμ or millimicron).

   1Å   = 1×10^{-10} metres
             = 1×10^{-1} nm

   (b) **Wavenumber** ($\bar{\nu}$) = number of waves per cm of path, i.e.

$$\frac{1}{\lambda(\text{in cm})},$$

the units being cm^{-1} or K (Kaysers)

$$kK = \text{kilo Kaysers} = \frac{1}{\lambda(\text{in cm})} \times 1000$$

i.e. $1kK = 1000\ cm^{-1}$

(c) **Frequency** (ν) = number of waves per unit time—usually per s
= $c/\lambda = 3 \times 10^{10}/\lambda(\text{in cm})$ (c = velocity of light in cm/s).

Instruments

Spectrometers have three essential constituents:

(1) *Light Source*

Hydrogen or Deuterium Arc Lamps for the U.V.,
Tungsten Filament Lamps for the visible wavelengths.

(2) *A Monochromator*

(3) *Photometer*

Light from the selected lamps is 'dispersed' by the prism or diffraction grating of the Monochromator into a 'spectrum'; light leaves the prism or grating at different angles each characteristic for a particular wavelength. The prism or grating can be rotated by a calibrated dial so that light of a particular wavelength passes out of the Monochromator through an exit slit into the sample compartment and thence to the photometer.

The photometer used is one which will readily compare light intensities such as I_0 and I of the equations above. Without special calibration it is not possible to measure light intensity in absolute terms with these photometers. In Single Beam Manual Spectrophotometers, such as the Unicam SP 500 and the Hilger 'Uvispeck', the output of the photometer is balanced against a stable reference voltage without the sample in the light beam. The sample is then placed in the light beam and only a fraction of the reference voltage is needed to balance the photometer output. This fraction corresponds to a certain value of D and the corresponding value of per cent Transmission. It is measured by a calibrated potentiometer across which the reference voltage is applied. Balance is indicated by a small electrical meter.

Because of the variation with wavelength of light energy available from the source, the proportion of light lost by scattering and absorption in the optical components, and the sensitivity of detectors, it is necessary to set the 0 extinction (100% transmission) position of the Photometer for each wavelength at which measurements are made.

In Double-Beam instruments such as the Optica CF 4 DR spectrophotometer, light from the Monochromator is automatically passed alternately through the reference and sample materials. The relative amounts of light passing to the detector (photometer) by the two paths is determined by comparing the size of the two electrical signals.

Electromagnetic Radiation

	Wavelength range		Wave number range
	nm	μm	cm^{-1}
X-rays	1–100	–	
Ultra violet	100–400	0.1–0.4	100 000–25 000
Violet	400–450	0.4–0.45	25 000–22 200
Blue	450–500	0.45–0.50	22 200–20,000
Green	500–575	0.50–0.575	20 000–17 400
Yellow	575–590	0.575–0.59	17 400–16 950
Orange	590–650	0.59–0.65	16 950–15 400
Red	650–750	0.65–	15 400–13 330
Near infra-red	750–1000	0.75–1.00	13 330–10 000
Far infra-red	1000–16 000	1.00–16.00	10 000–625

Relationship of Transmission to Extinction

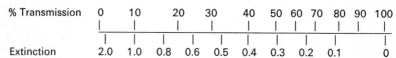

% Transmission	0	10	20	30	40	50	60	70	80	90	100
Extinction	2.0	1.0	0.8	0.6	0.5	0.4	0.3	0.2	0.1		0

Glossary

Absorption Band A region of the absorption spectrum in which the absorptivity passes through a maximum or inflection. This term usually refers to all the absorption due to a single electronic transition, including all vibrational sub-bands.

Absorption Maximum The wavelength (λ_{max}) at which a peak occurs in the absorption curve. There may be, and generally are, several such maxima.

Absorptivity or ***Specific Absorption Coefficient*** (k or α) The absorbance divided by the product of concentration and path length. It is absorbance per unit concentration and thickness, i.e., the *Specific* absorbance. The concentration may be expressed in moles per litre (Molar Extinction Coefficient), or in percentage weight/volume (Specific Extinction Coefficient).

Franck–Condon Principle During an electronic transition the position and momentum of nuclei remain unchanged due to the greater speed of electronic motion relative to nuclear motion.

Half-Band Width The width of a particular absorption band at points where $\alpha = \frac{1}{2}_{max}$, usually expressed in reciprocal centimetres, cm^{-1}, and indicated as $\Delta\nu$. This value is particularly useful in the comparison of the intensities of similar bands in the spectra of related compounds, since approximately,

$$\int \alpha \, d\nu \propto \alpha_{max} \, \Delta\nu$$

Optical density A term which is becoming obsolete and replaced by absorbance, with which it is identical.

Transmittance (T) The ratio of the intensity of the light transmitted by a sample (I) to that incident on the sample (I_0) both measured under identical conditions.

Vibrational structure The sub-bands associated with a single electronic transition and arising from the small energy difference of various vibrational levels of the molecules in both ground and excited states.

Wavelength (λ) The distance, measured along the line of propagation, between two points on adjacent waves which are in phase.

Wavenumber (ν) The number of waves per unit length in a vacuum, the reciprocal of wavelength, λ, usually measured in reciprocal centimetres, cm^{-1}.

References

1. ALLEN, M. B., ARNON, D. I., CAPINDALE, J. B., WHATLEY, F. R. and DURHAM, L. J. (1955). Photosynthesis by isolated chloroplasts. III. Evidence for complete photosynthesis. *J. Am. chem. Soc.*, **77**, 4149–4155.
2. ALLEN, M. B., PIETTE, L. R. and MURCHIO, J. C. (1962). Free radicals in photosynthetic reactions. I. Electron paramagnetic resonance signals from illuminated *Chlorella pyrenoidosa*. *Biochim. biophys. Acta*, **60**, 539–547.
3. AMESZ, J. and FORK, D. C. (1967). Quenching of chlorophyll fluorescence by quinones in algae and chloroplasts. *Biochim. biophys. Acta*, **143**, 97–107.
4. ANDERSON, J. M. and BOARDMAN, N. K. (1966). Fractionation of the photochemical systems of photosynthesis. I. Chlorophyll contents and photochemical activities of particles isolated from spinach chloroplasts. *Biochim. biophys. Acta*, **112**, 403–421.
5. ARNON, D. I. (1967). Photosynthetic activity of isolated chloroplasts. *Physiol. Rev.*, **47**, 317–358.
6. ARNON, D. I., WHATLEY, F. R. and ALLEN, M. B. (1954). Photosynthesis by isolated chloroplasts. II. Photosynthetic phosphorylation, the conversion of light into phosphate bond energy. *J. Am. chem. Soc.*, **76**, 6324–6329.
7. AVRON, M. and SHAVIT, N. (1965). Inhibitors and uncouplers of photophosphorylation. *Biochim. biophys. Acta*, **109**, 317–331.
8. BAMBERGER, E. S. and GIBBS, M. (1965). Effect of phosphorylated compounds and inhibitors on CO_2 fixation by intact spinach chloroplasts. *Pl. Physiol., Lancaster*, **40**, 919–926.
9. BASSHAM, J. A. (1964). Kinetic studies of the photosynthetic carbon reduction cycle. *A. Rev. Pl. Physiol.*, **15**, 101–120.
10. BENDALL, D. S. and HILL, R. (1968). Haem-proteins in photosynthesis. *A. Rev. Pl. Physiol.*, **19**, 167–186.
11. BENDER, M. M. (1968). Mass spectrometric studies of carbon 13 variations in corn and other grasses. *Am. J. Sci. Radiocarbon Suppl.*, **10**, 468–472.

116   REFERENCES

12. BENSON, A. A. and CALVIN, M. (1950). The path of carbon in photosynthesis, VII. Respiration and photosynthesis. *J. exp. Bot.*, **1**, 63–68.
13. BLACKMAN, F. F. and SMITH, A. M. (1911). Experimental researches on vegetable assimilation and respiration. IX. On assimilation in submerged water-plants and its relation to the concentration of carbon dioxide and other factors. *Proc. R. Soc.*, *B*, **83**, 389–412.
14. BRADBEER, J. W. and RACKER, E. (1961). Glycollate formation from fructose-6-phosphate by cell-free preparations. *Fedn Proc. Fedn Am. Socs exp. Biol.*, **20**, A–88d.
15. BRODY, S. S. and RABINOWITCH, E. (1957). Excitation lifetime of photosynthetic pigments *in vitro* and *in vivo*. *Science, N.Y.*, **125**, 555.
16. BROWN, A. H. (1953). The effects of light on respiration using isotopically enriched oxygen. *Am. J. Bot.*, **40**, 719–729.
17. BROWN, H. T. and ESCOMBE, F. (1905). Researches on some of the physiological processes of green leaves, with specific reference to the interchange of energy between the leaf and its surroundings. *Proc. R. Soc.*, *B*, **76**, 29–111.
18. BROWN, J. S. and FRENCH, C. S. (1959). Absorption spectra and relative photostability of the different forms of chlorophyll in *Chlorella*. *Pl. Physiol., Lancaster*, **34**, 305–309.
19. BULL, T. A. (1969). Photosynthetic efficiencies and photorespiration in Calvin cycle and C_4 dicarboxylic acid plants. *Crop Sci.*, **9**, 726–729.
20. BURSTRÖM, H. (1945). Photosynthesis and assimilation of nitrate by wheat leaves. *A. Rep. Agric. College Sweden*, 11.
21. BUTLER, W. L. (1963). Effect of light intensity on the far-red inhibition of chlorophyll *a* fluorescence *in vivo*. *Biochim. biophys. Acta*, **66**, 275–276.
22. CALVIN, M. and PON, N. G. (1959). Carboxylation and decarboxylations. *Symposium on Enzyme Reaction Mechanisms*. Oakridge, Nat. Lab., 51–74.
23. CHANCE, B. (1952). Spectra and reaction kinetics of respiratory pigments of homogenized and intact cells. *Nature, Lond.*, **169**, 215–221.
24. CHANCE, B. (1954). Spectrophotometry of intracellular respiratory pigments. *Science, N.Y.*, **120**, 767–775.
25. CLAYTON, R. K. (1963). Toward the isolation of a photochemical reaction center in *Rhodopseudomonas spheroides*. *Biochim. biophys. Acta*, **75**, 312–323.
26. COOMBS, J. and WHITTINGHAM, C. P. (1966). The mechanism of inhibition of photosynthesis by high partial pressures of oxygen in *Chlorella*. *Proc. R. Soc.*, *B*, **164**, 511–520.
27. DAS, M. and GOVINDJEE (1967). A long-wave absorbing form of chlorophyll *a* responsible for the 'red drop' in fluorescence at 298K and the F_{723} band at 77K. *Biochim. biophys. Acta*, **143**, 570–576.
28. DAVENPORT, H. E. and HILL, R. (1952). The preparation and some properties of cytochrome *f*. *Proc. R. Soc.*, *B*, **139**, 327–345.
29. DECKER, J. P. (1955). A rapid postillumination deceleration of respiration in green leaves. *Pl. Physiol., Lancaster*, **30**, 82–84.
30. DE DUVE, C. and BAUDHUIN, P. (1966). Peroxisomes (microbodies and related particles). *Physiol. Rev.*, **46**, 323–357.
31. DILLEY, R. A. and VERNON, L. P. (1965). Ion and water transport processes related to the light-dependent shrinkage of spinach chloroplasts. *Archs Biochem. Biophys.*, **111**, 365–375.

32. DOLE, M. and JENKS, G. (1944). Isotopic composition of photosynthetic oxygen. *Science, N.Y.*, **100**, 409.

33. DUYSENS, L. N. M. (1952). Ph.D. thesis, Univ. Utrecht.

34. DUYSENS, L. N. M. (1954). Reversible changes in the absorption spectrum of *Chlorella* upon irradiation. *Science, N.Y.*, **120**, 353–354.

35. DUYSENS, L. N. M., AMESZ, J. and KAMP, B. M. (1961). Two photochemical systems in photosynthesis. *Nature, Lond.*, **190**, 510–511.

36. DUYSENS, L. N. M. and SWEERS, H. E. (1963). Mechanism of two photochemical reactions in algae as studied by means of fluorescence. In *Studies on Microalgae and Photosynthetic Bacteria*. U. Tokyo Press, 353–372.

37. EL-SHARKAWY, M. A., LOOMIS, R. S. and WILLIAMS, W. A. (1967). Apparent reassimilation of respiratory carbon dioxide by different plant species. *Physiologia Pl.*, **20**, 171–186.

38. EMERSON, R. and ARNOLD, W. (1933). The photochemical reaction in photosynthesis. *J. gen. Physiol.*, **16**, 191–205.

39. EMERSON, R., CHALMERS, R. and CEDERSTRAND, C. (1957). Some factors influencing the long-wave limit of photosynthesis. *Proc. natn. Acad. Sci.*, *U.S.A.*, **43**, 133–143.

40. EMERSON, R. and LEWIS, C. M. (1943). The dependence of the quantum yield of *Chlorella* photosynthesis on wave length of light. *Am. J. Bot.*, **30**, 165–178.

41. EVERSON, R. G., COCKBURN, W. and GIBBS, M. (1967). Sucrose as a product of photosynthesis in isolated spinach chloroplasts. *Pl. Physiol.*, *Lancaster*, **42**, 840–844.

42. FOCK, H., SCHAUB, H., HILGENBERG, W. and EGLE, K. (1969). Über den Einfluss niedriger und hoher O_2-Partialdrucke auf den Sauerstoff und Kohlendioxidumsatz von *Amaranthus* und *Phaseolus* während der Lichtphase. *Planta*, **86**, 77–83.

43. FORRESTER, M. L., KROTKOV, G. and NELSON, C. D. (1966). Effect of oxygen on photosynthesis, photorespiration and respiration in detached leaves. I. Soybean. *Pl. Physiol.*, *Lancaster*, **41**, 422–431.

44. FREDERICK, S. E. and NEWCOMB, E. H. (1969). Microbody-like organelles in leaf cells. *Science, N.Y.*, **163**, 1353–1355.

45. FRENKEL, A. W. (1956). Photophosphorylation of adenine nucleotides by cell-free preparations of purple bacteria. *J. biol. Chem.*, **222**, 823–834.

46. FULLER, R. C. and NUGENT, N. A. (1969). Pteridines and the function of the photosynthetic reaction centre. *Proc. natn. Acad. Sci.*, **63**, 1311–1318.

47. GAFFRON, H. (1933). Über den Stoffwechsel der schwefelfreien Purpurbakterien. *Biochem. Z.*, **260**, 1–17.

48. GAFFRON, H. (1935). Über den Stoffwechsel der Purpurbakterien. II. *Biochem. Z.*, **275**, 301–319.

49. GAFFRON, H. and RUBIN, J. (1943). Fermentative and photochemical production of hydrogen in algae. *J. gen. Physiol.*, **26**, 219–240.

50. GEST, H. (1966). Comparative biochemistry of photosynthetic processes. *Nature, Lond.*, **209**, 879–882.

51. GIBBS, M. and CYNKIN, M. A. (1958). Conversion of carbon-14 dioxide to starch glucose during photosynthesis by spinach chloroplasts. *Nature, Lond.*, **182**, 1241–1242.

52. GIBBS, M., LATZKO, E., O'NEALE, D. and HEW, C.-S. (1970). Photosynthetic carbon fixation by isolated maize chloroplasts. *Biochem. biophys. Res. Commun.*, **40**, 1356–1361.

53. GOLDSWORTHY, A. (1966). Experiments on the origin of CO_2 released by tobacco leaf segments in the light. *Phytochem. Newsl.*, 5, 1013–1019.
54. GOLDSWORTHY, A. (1968). Comparison of the kinetics of photosynthetic carbon dioxide fixation in maize, sugar cane and tobacco, and its relation to photorespiration. *Nature, Lond.*, 217, 62.
55. HATCH, M. D. and SLACK, C. R. (1966). Photosynthesis by sugarcane leaves. A new carboxylation reaction and the pathway of sugar formation. *Biochem. J.*, 101, 103–111.
56. HATCH, M. D. and SLACK, C. R. (1970). Photosynthetic CO_2-fixation pathways. *A. Rev. Pl. Physiol.*, 21, 141–162.
57. HAXO, F. T. and BLINKS, L. R. (1950). Photosynthetic action spectra of marine algae. *J. gen. Physiol.*, 33, 389–422.
58. HEBER, U., PON, N. G. and HEBER, M. (1963). Localization of carboxydismutase and triosephosphate dehydrogenases in chloroplasts. *Pl. Physiol., Lancaster*, 38, 355–360.
59. HILL, R. (1937). Oxygen evolved by isolated chloroplasts. *Nature, Lond.*, 139, 881–882.
60. HILL, R. (1939). Oxygen produced by isolated chloroplasts. *Proc. R. Soc., B*, 127, 192–210.
61. HILL, R. and BENDALL, F. (1960). Function of the two cytochrome components in chloroplasts: a working hypothesis. *Nature, Lond.*, 186, 136–137.
62. HILL, R. and WALKER, D. A. (1959) Pyocyanine and phosphorylation with chloroplasts. *Pl. Physiol., Lancaster*, 34, 240–245.
63. HIND, G. and JAGENDORF, A. T. (1963). Separation of light and dark stages in photophosphorylation. *Proc. natn. Acad. Sci., U.S.A.*, 49, 715–722.
64. HIND, G. and JAGENDORF, A. T. (1965). Light scattering changes associated with the production of a possible intermediate in photophosphorylation. *J. biol. Chem.*, 240, 3195–3201.
65. HIND, G. and OLSON, J. M. (1968). Electron transport pathways in photosynthesis. *A. Rev. Pl. Physiol.*, 19, 249–282.
66. HOCH, G. and KOK, B. (1963). A mass spectrometer inlet system for sampling gases dissolved in liquid phases. *Archs Biochem. Biophys.*, 101, 160–170.
67. HORECKER, B. L., SMYRNIOTIS, P. Z. and KLENOW, H. (1953). The formation of sedoheptulose phosphate from pentose phosphate. *J. biol. Chem.*, 205, 661–682.
68. IZAWA, S. and GOOD, N. E. (1965). The number of sites sensitive to 3-(3,4-dichlorophenyl)-1,1-dimethylurea,   3-(4-chlorophenyl)-1,1-dimethylurea and 2-chloro-4-(2-propylamino)-6-ethylamine-*s*- triazine in isolated chloroplasts. *Biochim. biophys. Acta*, 102, 20–38.
69. JACOBI, G. and LEHMANN, H. (1968). Die Fragmentation isolierter Chloroplasten. I. Die funktionelle und strukturelle Beurteilung von fragmenten ultrascholl-behandelter Chloroplasten. *Z. Pflanzenphysiol.*, 59, 457–476.
70. JAGENDORF, A. T. and AVRON, M. (1959). Inhibitors of photosynthetic phosphorylation in relation to electron and oxygen transport pathways of chloroplasts. *Archs Biochem. Biophys.*, 80, 246–257.
71. JENSEN, R. G. (1971). Activation of CO_2 fixation in isolated spinach chloroplasts. *Biochim. biophys. Acta*, 234, 360–370.
72. JENSEN, R. G. and BASSHAM, J. A. (1966). Photosynthesis by isolated chloroplasts. *Proc. natn. Acad. Sci., U.S.A.*, 56, 1091–1101.

73. JOLIOT, P. (1965). Cinétiques des réactions liées a l'émission d'oxygène photosynthétique. *Biochim. biophys. Acta*, **102**, 116–134.
74. JOLIOT, P. (1968). Kinetic studies of photosystem II in photosynthesis. *Photochem. photobiol.*, **8**, 451–463.
75. KAMEN, M. D. and GEST, H. (1949). Evidence for a nitrogenase system in the photosynthetic bacterium *Rhodospirillum rubrum*. *Science, N.Y.* **109**, 560.
76. KANDLER, O. and GIBBS, M. (1956). Asymmetric distribution of C^{14} in the glucose phosphates formed during photosynthesis. *Pl. Physiol., Lancaster*, **31**, 411–412.
77. KANDLER, O. and TANNER, W. (1966). Die Photoassimilation von Glucose als Indikator für die Lichtphosphorylierung *in vivo*. *Ber. dt. bot. Ges.*, **79**, (48)–(57).
78. KATOH, S. (1960). A new copper protein from *Chlorella ellipsoidea*. *Nature, Lond.*, **186**, 533–534.
79. KEYS, A. J. (1968). The intracellular distribution of free nucleotides in the tobacco leaf. Formation of adenosine 5'-phosphate from adenosine 5'-triphosphate in the chloroplasts. *Biochem. J.*, **108**, 1–8.
80. KNAFF, D. B. and ARNON, D. I. (1969). A concept of three light reactions in photosynthesis by green plants. *Proc. natn. Acad. Sci., U.S.A.*, **64**, 715–722.
81. KOK, B. (1957). Light induced absorption changes in photosynthetic organisms. *Acta bot. neerl.*, **6**, 316–336.
82. KOK, B. (1961). Partial purification and determination of oxidation reduction potential of the photosynthetic chlorophyll complex absorbing at 700 mµ. *Biochim. biophys. Acta*, **48**, 527–533.
83. KOK, B. and BEINERT, H. (1962). The light induced EPR signal of photocatalyst P_{700}. II. Two light effects. *Biochem. biophys. Res. Commun.*, **9**, 349–354.
84. KOK, B. and RURAINSKI, H. J. (1965). Plastocyanin photo-oxidation by detergent-treated chloroplasts. *Biochim. biophys. Acta*, **94**, 588–590.
85. KORTSCHAK, H. P., HARTT, C. E. and BURR, G. O. (1965). Carbon dioxide fixation in sugarcane leaves. *Pl. Physiol., Lancaster*, **40**, 209–213.
86. KREY, A. and GOVINDJEE (1966). Fluorescence studies on a red alga, *Porphyridium cruentum*. *Biochim. biophys. Acta*, **120**, 1–18.
87. KROTKOV, G. (1963). Effect of light on respiration. In *Photosynthetic Mechanisms of Green Plants*. Publ. 1145, Nat. Acad. Sci., Washington, D.C., 452–454.
88. LATIMER, P., BANNISTER, T. T. and RABINOWITCH, E. (1956). Quantum yields of fluorescence of plant pigments. *Science, N.Y.*, **124**, 585–586.
89. LATIMER, P., BANNISTER, T. T. and RABINOWITCH, E. (1957). The absolute quantum yields of fluorescence of photosynthetically active pigments. In *Research in Photosynthesis*. Interscience, New York, 107–112.
90. LAVOREL, J. (1962). Hétérogénéité de la chlorophylle *in vivo*. I. Spectres d'émission de fluorescence. *Biochim. biophys. Acta*, **60**, 510–523.
91. LAVOREL, J. (1964). Hétérogénéité de la chlorophylle *in vivo*. II. Polarisation et spectres d'action de fluorescence. *Biochim. biophys. Acta*, **88**, 20–36.
92. LESTER, R. J. and CRANE, F. L. (1959). The natural occurrence of co-enzyme Q and related compounds. *J. biol. Chem.*, **234**, 2169–2175.
93. LEVINE, R. P. (1969). The analysis of photosynthesis using mutant strains of algae and higher plants. *A. Rev. Pl. Physiol.*, **20**, 523–540.

94. LEVINE, R. P. and SMILLIE, R. M. (1962). The pathway of triphosphopyridine nucleotide photoreduction in *Chlamydomonas reinhardi*. *Proc. natn. Acad. Sci.*, *U.S.A.*, **48**, 417–421.

95. LEVINE, R. P. and SMILLIE, R. M. (1963). The photosynthetic electron transport chain of *Chlamydomonas reinhardi*. I. Triphosphopyridine nucleotide photoreduction in wild-type and mutant strains. *J. biol. Chem.*, **238**, 4052–4057.

96. MACLACHLAN, G. A. and PORTER, H. K. (1959). Replacement of oxidation by light as the energy source for glucose metabolism in tobacco leaf. *Proc. R. Soc.*, B, **150**, 460–473.

97. MACROBBIE, E. A. C. (1965). The nature of the coupling between light energy and active ion transport in *Nitella translucens*. *Biochim. biophys. Acta*, **94**, 64–73.

98. MAYNE, B. C. and CLAYTON, R. K. (1966). Luminescence of chlorophyll in spinach chloroplasts induced by acid-base transition. *Proc. natn. Acad. Sci.*, *U.S.A.*, **55**, 494–497.

99. MICHEL, J. M. and MICHEL-WOLWERTZ, M. R. (1970). Fractionation and photochemical activities of photosystems isolated from broken spinach chloroplasts by sucrose-density gradient centrifugation. *Photosynthetica*, **4**, 146–155.

100. MITCHELL, P. (1966). Chemiosmotic coupling in oxidative and photosynthetic phosphorylation. *Biol. Rev.*, **41**, 445–502.

101. MONTEITH, J. L., SZEICZ, G. and YABUKI, K. (1964). Crop photosynthesis and the flux of carbon dioxide below the canopy. *J. appl. Ecol.*, **1**, 321–337.

102. MORTENSON, L. E., VALENTINE, R. C. and CARNAHAN, J. E. (1962). An electron transport factor from *Clostridium pasteurianum*. *Biochem. biophys. Res. Commun.*, **7**, 448–452.

103. MOSS, D. N. (1962). The limiting carbon dioxide concentration for photosynthesis. *Nature, Lond.*, **193**, 587.

104. MULLER, A., RUMBERG, B. and WITT, H. T. (1963). On the mechanism of photosynthesis. *Proc. R. Soc.*, B, **157**, 313–332.

105. MÜLLHOFER, G. and ROSE, I. A. (1965). The position of carbon-carbon bond cleavage in the ribulose diphosphate carboxydismutase reaction. *J. biol. Chem.*, **240**, 1341–1346.

106. MURTY, N. R. and RABINOWITCH, E. (1965). Fluorescence decay studies of chlorophyll *a in vivo*. *Biophys. J.*, **5**, 655–661.

107. MYERS, J. (1971). Enhancement studies in photosynthesis. *A. Rev. Pl. Physiol.*, **22**, 289–312.

108. MYERS, J. and FRENCH, C. S. (1960). Evidences from action spectra for a specific participation of chlorophyll *b* in photosynthesis. *J. gen. Physiol.*, **43**, 723–736.

109. PACKER, L. and SIEGENTHALER, P.–A. (1965). Light-dependent volume changes and reactions in chloroplasts. II. Action of anions. *Pl. Physiol., Lancaster*, **40**, 1080–1085.

110. PARSON, W. W. (1967). Flash-induced absorbance changes in *Rhodospirillum rubrum* chromatophores. *Biochim. biophys. Acta*, **131**, 154–172.

111. PEDERSEN, T. A., KIRK, M. and BASSHAM, J. A. (1966). Light-dark transients in levels of intermediate compounds during photosynthesis in air-adapted *Chlorella*. *Physiologia Pl.*, **19**, 219–231.

112. PENMAN, H. L. and SCHOFIELD, R. K. (1951). Some physical aspects of assimilation and transpiration. *Symp. Soc. exp. Biol.*, **5**, 115–129.

113. QUAYLE, J. R., FULLER, R. C., BENSON, A. A. and CALVIN, M. (1954). Enzymatic carboxylation of ribulose diphosphate. *J. Am. chem. Soc.*, **76**, 3610–3611.

114. ROBERTS, G. R., KEYS, A. J. and WHITTINGHAM, C. P. (1970). The transport of photosynthetic products from the chloroplasts of tobacco leaves. *J. exp. Bot.*, **21**, 683–692.

115. ROHR, W. and BASSHAM, J. A. (1964). Two-dimensional high-voltage, low-temperature paper electrophoresis of ^{14}C-labelled products of photosynthesis with $^{14}CO_2$. *Analyt. Biochem.*, **9**, 343–350.

116. RUBEN, S., HASSID, W. Z. and KAMEN, W. D. (1939). Radioactive carbon in the study of photosynthesis. *J. Am. chem. Soc.*, **61**, 661–663.

117. SAN PIETRO, A. and LANG, H. M. (1958). Photosynthetic pyridine nucleotide reductase. I. Partial purification and properties of the enzyme from spinach. *J. biol. Chem.*, **231**, 211–229.

118. SCHWARTZ, M. (1967). Wavelength-dependent quantum yield of ATP synthesis and NADP reduction in normal and dichlorodimethylphenylurea-poisoned chloroplast. *Biochim. biophys. Acta*, **131**, 559–570.

119. SHIN, M. and ARNON, D. I. (1965). Enzymic mechanisms of pyridine nucleotide reduction in chloroplasts. *J. biol. Chem.*, **240**, 1405–1411.

120. SLATYER, R. O. and BIERHUIZEN, J. F. (1964). The influence of several transpiration suppressants on transpiration, photosynthesis and water use efficiency of cotton leaves. *Aust. J. biol. Sci.*, **17**, 131–146.

121. SMITH, J. H. C. (1949). Products of photosynthesis. In *Photosynthesis in Plants*. Iowa State College Press, 53–94.

122. SMITH, J. H. C. (1961). Some physical and chemical properties of the protochlorophyll holochrome. In *Biological Structure and Function*. Proc. 1st IUB/IUBS Int. Symp., Academic Press, **11**, 325–338.

123. STANIER, R. Y. (1961). Photosynthetic mechanisms in bacteria and plants: Development of a unitary concept. *Bact. Rev.*, **25**, 1–17.

124. STILLER, M. (1962). The path of carbon in photosynthesis. *A. Rev. Pl. Physiol.*, **13**, 151–170.

125. STOCKING, C. R. (1959). Chloroplast isolation in non-aqueous media. *Pl. Physiol., Lancaster*, **34**, 56–61.

126. STOCKING, C. R. and LARSON, S. (1969). A chloroplast cytoplasmic shuttle and the reduction of extraplastid NAD. *Biochem. biophys. Res. Commun.*, **37**, 278–282.

127. STREHLER, B. L. and ARNOLD, W. (1951). Light production by green plants. *J. gen. Physiol.*, **34**, 809–820.

128. SYRETT, P. J. (1966). The kinetics of isocitrate lyase formation in *Chlorella*: Evidence for the promotion of enzyme synthesis by photophosphorylation. *J. exp. Bot.*, **17**, 641–654.

129. TEALE, F. W. J. (1960). The extent of energy migration and chlorophyll *a* orientation in *Chlorella*. *Biochim. biophys. Acta*, **42**, 69–75.

130. THOMPSON, C. M. and WHITTINGHAM, C. P. (1968). Glycollate metabolism in photosynthesising tissue. *Biochim. biophys. Acta*, **153**, 260–269.

131. TOLBERT, N. E. (1958). Secretion of glycollic acid by chloroplasts. In *The Photochemical Apparatus—its Structure and Function*. Brookhaven Symposia in Biology, No. 11, 271–275.

132. TOLBERT, N. E., OESER, A., YAMAZAKI, R. K., HAGEMAN, R. H. and KISAKI, T. (1969). A survey of plants for leaf peroxisomes. *Pl. Physiol., Lancaster*, **44**, 135–147.

133. TOLBERT, N. E. and ZILL, L. P. (1956). Excretion of glycollic acid by algae during photosynthesis. *J. biol. Chem.*, **222**, 895–906.
134. TREBST, A. and FIEDLER, F. (1962). Über die Ursache der asymmetrischen C-Verteilung in der Hexose bei der Photosynthese mit Chloroplasten. *Z. Naturf., B*, **17**, 553–558.
135. TREGUNNA, E. B., KROTKOV, G. and NELSON, C. D. (1966). Effect of oxygen on the rate of photorespiration in detached tobacco leaves. *Physiologia Pl.*, **19**, 723–733.
136. VAN NIEL, C. B. (1941). The bacterial photosyntheses and their importance for the general problem of photosynthesis. *Adv. Enzymol.*, **1**, 263–328.
137. VERNON, L. P. and ZAUGG, W. S. (1960). Photoreductions by fresh and aged chloroplasts: Requirement for ascorbate and 2,6-dichlorophenolindophenol with aged chloroplasts. *J. biol. Chem.*, **235**, 2728–2733.
138. VREDENBERG, W. J. and DUYSENS, L. N. M. (1963). Transfer of energy from bacteriochlorophyll to a reaction centre during bacterial photosynthesis. *Nature. Lond.*, **197**, 355–357.
139. WALKER, D. A. (1965). Correlation between photosynthetic activity and membrane integrity in isolated pea chloroplasts. *Pl. Physiol., Lancaster*, **40**, 1157–1161.
140. WALKER, D. A. and HILL, R. (1967). The relation of oxygen evolution to carbon assimilation with isolated chloroplasts. *Biochim. biophys. Acta*, **131**, 330–338.
141. WARBURG, O. (1919). Über die Geschwindigkeit der photochemischen Kohlensaurezersetzung in lebenden Zellen. *Biochem. Z.*, **100**, 230–270.
142. WEBER, G. and TEALE, F. W. J. (1957). Determination of the absolute quantum yield of fluorescent solutions. *Trans. Faraday Soc.*, **53**, 646–655.
143. WESTLAKE, D. F. (1963). Comparisons of plant productivity. *Biol. Rev.*, **38**, 385–425.
144. WHATLEY, F. R., TAGAWA, K. and ARNON, D. I. (1963). Separation of the light and dark reactions in electron transfer during photosynthesis. *Proc. natn. Acad. Sci., U.S.A.*, **49**, 266–270.
145. WHITTINGHAM, C. P. and PRITCHARD, G. G. (1963). The production of glycollate during photosynthesis in *Chlorella*. *Proc. R. Soc., B*, **157**, 366–380.
146. WISHNICK, M. and LANE, M. D. (1969). Inhibition of ribulose diphosphate carboxylase by cyanide. Inactive ternary complex of enzyme, ribulose diphosphate, and cyanide. *J. biol. Chem.*, **244**, 55–59.
147. WITT, H. T. (1955). Kurzzeitige Absorptionsänderungen beim Primarprozess der Photosynthese. *Naturwissenschaften*, **42**, 72–73.
148. YOCUM, C. F. and SAN PIETRO, A. (1970). Ferredoxin-reducing substance (FRS) from spinach. II. Separation and assay. *Archs Biochem. Biophys.*, **140**, 152–157.
149. ZELITCH, I. (1959). The relationship of glycolic acid to respiration and photosynthesis in tobacco leaves. *J. biol. Chem.*, **234**, 3077–3081.

Index

absorption of light, 56 *et seq.*, 109
acid–base phosphorylation, 106
action spectra, 68
active glycoaldehyde, 27
algal mutants, 83, 99
Amaranthus, 38
assimilatory quotient, 3
Athiorhodaceae, 85, 92
atomic absorption spectra, 56
Atriplex, 38, 43

bacteriochlorophyll, 85
  in vivo, 101
barley, photosynthesis by, 17
bicarbonate as carbon source, 2
Blackman reaction, 15
bundle sheath, 29

C^4-plants, 28, 43
Calvin photosynthetic cycle, 16 *et seq.*
carbon dioxide, burst, 41
  concentration and rate of photosynthesis, 10
  fixation by chloroplasts, 47
  tracer, 16

carbon, isotope ratios, 43
  isotopes as tracers, 16
  metabolism, 16 *et seq.*
carboxylation and photosynthesis, 24, 27, 31
carotenoids, 58, 66, 75
chemiluminescence of chlorophyll, 107
Chlamydomonas mutants, 83, 99
Chloraceae, 85
Chlorella, action spectrum of, 68
  fluorescence spectrum of, 71
  glucose assimilation in, 91
  tracer studies with, 17, 33–37
chlorophyll, absorption bands, 58
  chemiluminescence, 107
  energy transfer in, 66
  fluorescence, 59, 65
  phosphorescence, 59
chlorophyll forms *in vivo*, 62, 80
chloroplast, conformational changes 105
  distribution in leaf, 29
  fractionation, 102
  metabolism, 48, 50
compensation point, 37

crop production, efficiency of, 2
cytochromes, absorption bands of, 73, 75
    role in photosynthesis, 79, 94, 100–107

dark reaction of photosynthesis, 15
DCMU as inhibitor, 65, 91, 98
difference spectra, 71
diffusion, and stomatal control, 9
    of carbon dioxide, 5–8
    of water vapour, 2, 5–8

electron transport in photosynthesis, 94 *et seq.*
Emerson effect, 77
energy levels of chlorophyll, 60
energy requirement of photosynthesis, 2
enhancement spectra, 77
equivalent air diffusion path of leaf, 8
etiolated plants, 64
excited states, 57 *et seq.*

fat production in photosynthesis, 4
ferredoxin, 88, 101
Fick's Law of diffusion, 5
flash spectroscopy, 74
fluorescence, chlorophyll, 60
    quencher, 65, 100
    sensitized, 66
    time course of, 60, 65
    yield of, 62, 65
fraction 1 protein, 25
free energy change, in Calvin cycle, 24
    in carboxylation, 24, 31
    in photosynthesis, 2

glycine decarboxylation, 34, 44
glycollic acid, biosynthesis of, 36
    formation in photosynthesis, 32
    metabolism of, 44

green sulphur bacteria, 85

Hatch and Slack cycle, 28
heats of combustion, 4
Hill reaction, 47, 86
hydrogen evolution, 87
hydroxymethanesulphonates, 32

inhibitors of photosynthesis, 15, 25, 65, 91, 98
isonicotinylhydrazide, 32

leaf anatomy, 2, 29
limiting factors, 14
line spectra, 57

maize, 39
malic acid, 28
malic enzyme, 31
maple, 39
mesophyll cell function in C^4-plants 30
Mitchell theory, 104
molecular absorption spectra, 58
    degradation of intermediates, 18, 26, 36
mutants, 83, 99

Nitella, 92
nitrogen fixation, 87
nuclear magnetic resonance spectra, 81

oxygen, effect on photorespiration, 43
    effect on photosynthetic products, 37
    effect on rate of photosynthesis, 41
    isotopes of, 4, 39

P_{700}, 75, 83, 100
peroxisomes, 44

phosphoenolpyruvic acid, 28, 30
phosphoglyceric acid as first inter-
    mediate, 18
phosphorescence of chlorophyll, 61
phosphorus isotopes, 53
photophosphorylation, 90
    in vivo, 91
    mechanism of, 103
photoreduction, 87
photorespiration, 32 *et seq.*
    and phosphorylation, 46
    enzymes of, 44
    site of, 44
    substrate of, 43
photosynthetic bacteria, 84 *et seq.*,
    96
    pigments of, 77
photosynthetic reaction centres, 107
    reduction cycle, 21
phycobilins, 59, 67, 69, 80
plastocyanin, 98
plastoquinone, 65, 75, 99
Porphyridium, 74, 79, 80
products of photosynthesis, 3, 17,
    34
    in chloroplasts, 50
protein production in photosynthe-
    sis, 4
protochlorophyll, 64

quantum energy, 56

reaction centres, 69, 107
red drop, 71, 77
resonance transfer of energy, 67

serine, 34, 44
short-lived intermediate, 27
shuttle mechanisms, 54
stoma, 2
sugar-cane, 28, 39
sugar interconversions in photo-
    synthesis, 24

time sequence in photosynthesis, 18
Thiorhodaceae, 85
tobacco, 39
tomato, 39
transfer between chloroplast and
    cytoplasm, 51
transient phenomena, 22
transketolase, 22
transpiration and photosynthesis, 5
    et seq.
triplet state, 61, 74
two photochemical reactions, 96

Van Niel's theory of photosynthesis,
    85

water loss in photosynthesis, 8
wheat, 39